Modeling and Simulation
Linking Entertainment and Defense

Committee on Modeling and Simulation:
Opportunities for Collaboration Between the
Defense and Entertainment Research Communities

Computer Science and Telecommunications Board

Commission on Physical Sciences,
Mathematics, and Applications

National Research Council

NATIONAL ACADEMY PRESS
Washington, D.C. 1997

NOTICE: The project that is the subject of this report was approved by the Governing Board of the National Research Council, whose members are drawn from the councils of the National Academy of Sciences, the National Academy of Engineering, and the Institute of Medicine. The members of the committee responsible for the report were chosen for their special competences and with regard for appropriate balance.

This report has been reviewed by a group other than the authors according to procedures approved by a Report Review Committee consisting of members of the National Academy of Sciences, the National Academy of Engineering, and the Institute of Medicine.

The National Academy of Sciences is a private, nonprofit, self-perpetuating society of distinguished scholars engaged in scientific and engineering research, dedicated to the furtherance of science and technology and to their use for the general welfare. Upon the authority of the charter granted to it by the Congress in 1863, the Academy has a mandate that requires it to advise the federal government on scientific and technical matters. Dr. Bruce Alberts is president of the National Academy of Sciences.

The National Academy of Engineering was established in 1964, under the charter of the National Academy of Sciences, as a parallel organization of outstanding engineers. It is autonomous in its administration and in the selection of its members, sharing with the National Academy of Sciences the responsibility for advising the federal government. The National Academy of Engineering also sponsors engineering programs aimed at meeting national needs, encourages education and research, and recognizes the superior achievements of engineers. Dr. William A. Wulf is president of the National Academy of Engineering.

The Institute of Medicine was established in 1970 by the National Academy of Sciences to secure the services of eminent members of appropriate professions in the examination of policy matters pertaining to the health of the public. The Institute acts under the responsibility given to the National Academy of Sciences by its congressional charter to be an adviser to the federal government and, upon its own initiative, to identify issues of medical care, research, and education. Dr. Kenneth I. Shine is president of the Institute of Medicine.

The National Research Council was organized by the National Academy of Sciences in 1916 to associate the broad community of science and technology with the Academy's purposes of furthering knowledge and advising the federal government. Functioning in accordance with general policies determined by the Academy, the Council has become the principal operating agency of both the National Academy of Sciences and the National Academy of Engineering in providing services to the government, the public, and the scientific and engineering communities. The Council is administered jointly by both Academies and the Institute of Medicine. Dr. Bruce Alberts and Dr. William A. Wulf are chairman and vice-chairman, respectively, of the National Research Council.

Support for this project was provided by the Defense Modeling and Simulation Office through Subcontract 4843 from RGB Technology Inc. Any opinions, findings, conclusions, or recommendations expressed in this material are those of the authors and do not necessarily reflect the views of the sponsors.

Library of Congress Catalog Card Number 97-68732
International Standard Book Number 0-309-05842-2

Additional copies of this report are available from:

National Academy Press
2101 Constitution Avenue, NW
Box 285
Washington, DC 20055
800/624-6242
202/334-3313 (in the Washington Metropolitan Area)
http://www.nap.edu

Copyright 1997 by the National Academy of Sciences. All rights reserved.

Printed in the United States of America

**COMMITTEE ON MODELING AND SIMULATION:
OPPORTUNITIES FOR COLLABORATION BETWEEN THE
DEFENSE AND ENTERTAINMENT RESEARCH COMMUNITIES**

MICHAEL ZYDA, Naval Postgraduate School, *Chair*
DONNA COX, University of Illinois, Urbana-Champaign
WARREN KATZ, MäK Technologies
JOSHUA LARSON-MOGAL, Silicon Graphics Inc.
GILMAN LOUIE, Spectrum HoloByte Inc.
PAUL LYPACZEWSKI, Alias | Wavefront
RANDY PAUSCH, Carnegie Mellon University
ALEXANDER SINGER, Independent Producer/Director
JORDAN WEISMAN, Virtual World Entertainment Inc.

Staff

JERRY R. SHEEHAN, Study Director
LISA L. SHUM, Project Assistant
GLORIA BEMAH, Administrative Assistant (through November 1996)

COMPUTER SCIENCE AND TELECOMMUNICATIONS BOARD

DAVID D. CLARK, Massachusetts Institute of Technology, *Chair*
FRANCES E. ALLEN, IBM T.J. Watson Research Center
JEFF DOZIER, University of California at Santa Barbara
SUSAN L. GRAHAM, University of California at Berkeley
JAMES GRAY, Microsoft Corporation
BARBARA J. GROSZ, Harvard University
PATRICK HANRAHAN, Stanford University
JUDITH HEMPEL, University of California at San Francisco
DEBORAH A. JOSEPH, University of Wisconsin
BUTLER W. LAMPSON, Microsoft Corporation
EDWARD D. LAZOWSKA, University of Washington
BARBARA H. LISKOV, Massachusetts Institute of Technology
JOHN MAJOR, Qualcomm Inc.
ROBERT L. MARTIN, Lucent Technologies
DAVID G. MESSERSCHMITT, University of California at Berkeley
CHARLES L. SEITZ, Myricom Inc.
DONALD SIMBORG, KnowMed Systems Inc.
LESLIE L. VADASZ, Intel Corporation

MARJORY S. BLUMENTHAL, Director
HERBERT S. LIN, Senior Staff Officer
JERRY R. SHEEHAN, Staff Officer
JULIE CLYMAN LEE, Administrative Assistant
LISA L. SHUM, Project Assistant
SYNOD P. BOYD, Project Assistant

COMMISSION ON PHYSICAL SCIENCES, MATHEMATICS, AND APPLICATIONS

ROBERT J. HERMANN, United Technologies Corporation, *Co-chair*
W. CARL LINEBERGER, University of Colorado, *Co-chair*
PETER M. BANKS, Environmental Research Institute of Michigan
LAWRENCE D. BROWN, University of Pennsylvania
RONALD G. DOUGLAS, Texas A&M University
JOHN E. ESTES, University of California at Santa Barbara
L. LOUIS HEGEDUS, Elf Atochem North America Inc.
JOHN E. HOPCROFT, Cornell University
RHONDA J. HUGHES, Bryn Mawr College
SHIRLEY A. JACKSON, U.S. Nuclear Regulatory Commission
KENNETH H. KELLER, University of Minnesota
KENNETH I. KELLERMANN, National Radio Astronomy Observatory
MARGARET G. KIVELSON, University of California at Los Angeles
DANIEL KLEPPNER, Massachusetts Institute of Technology
JOHN KREICK, Sanders, a Lockheed Martin Company
MARSHA I. LESTER, University of Pennsylvania
THOMAS A. PRINCE, California Institute of Technology
NICHOLAS P. SAMIOS, Brookhaven National Laboratory
L.E. SCRIVEN, University of Minnesota
SHMUEL WINOGRAD, IBM T.J. Watson Research Center
CHARLES A. ZRAKET, MITRE Corporation (retired)

NORMAN METZGER, Executive Director

Preface

The entertainment industry and the U.S. Department of Defense (DOD)—though differing widely in their motivations, objectives, and cultures—share a growing interest in modeling and simulation. In entertainment, modeling and simulation technology is a key component of a $30 billion annual market for video games, location-based entertainment, theme parks, and films. In defense, modeling and simulation provides a cost-effective means of conducting joint training; developing new doctrine, tactics, and operational plans; assessing battlefield conditions; and evaluating new and upgraded systems.

Recognizing this synergy, DOD's Defense Modeling and Simulation Office (DMSO) asked the National Research Council's Computer Science and Telecommunications Board to convene a multidisciplinary committee to evaluate the extent to which the entertainment industry and DOD might be able to better leverage each other's capabilities in modeling and simulation technology and to identify potential areas for greater collaboration (see Appendix C for committee members' biographies). The committee met in June and August 1996 to plan a two-day workshop that was held in Irvine, California, in October 1996 (see Appendixes A and B for the workshop agenda and list of participants). It met again in November 1996 to discuss the results of the workshop and to plan the structure and format of this summary report.

The workshop brought together more than 50 representatives of the entertainment and defense research communities to discuss technical challenges facing the two industries, identify obstacles to successful sharing of technology and joint research, and suggest mechanisms for facilitating greater collaboration. Participants were drawn from the film, video game, location-based entertainment, and theme park industries; DOD; defense contractors; and universities. They included top executives and government program managers as well as engineers, film directors, researchers from industry and academia, and university faculty. Through a series of presentations on electronic storytelling, strategy and war gaming, experiential computing and virtual reality, networked simulation, and low-cost simulation hardware, the committee attempted to encourage dialogue among these diverse stakeholders and stimulate discussion of research areas of interest to both the entertainment and defense industries. Because the workshop represented one of the first formal attempts to bridge the gap between the entertainment and defense communities, the committee also hoped to encourage personal contacts between members of the two communities as a means of facilitating future collaboration. As such, the 1996 workshop should be seen as part of an ongoing process that may continue beyond this project and this report.

This report represents the committee's attempt to capture key themes of the workshop discussions. Chapter 1 provides an overview of the applications of modeling and simulation technology in the entertainment and defense industries and discusses the historical flows of technology between them. It also reviews the potential benefits to collaboration and outlines the underlying technologies of modeling and simulation in which collaboration may be possible. Chapter 2 identifies common technical needs of DOD and the entertainment industry, identifying and describing areas in which the entertainment and defense communities appear to have similar interests and in which collaboration, at some level, may be possible. Chapter 3 describes other issues that must be addressed in order to facilitate collaboration and sharing of research. These include the needs to develop the necessary human resources, establish mechanisms for information sharing and technology transfer, strengthen the research base, and overcome cultural differences between the two communities. As Chapter 3 notes, collaboration between the entertainment and defense research communities will require far more than a list of common research interests. Structures must be put in place to facilitate collaboration and to allow greater sharing of information between the two communities; differences in culture and business practices must be overcome, though not necessarily altered. Putting these elements in place will facilitate collaboration over time on an ever-changing set of common technologies and research areas.

This report benefited from the contributions of many people throughout the modeling and simulation community. Workshop participants, through their presentations and discussion, provided the committee with

PREFACE   ix

much of the material used in this report. The committee is especially grateful to those participants who submitted position papers outlining the research challenges in their particular fields of interest. The committee drew from these papers in preparing this report; the papers are reproduced in Appendix D. External reviewers of an early draft of this report also provided valuable comments.

Staff members of the U.S. Army's Topographic Engineering Center and Joint Precision Strike Demonstration provided the committee with an informative demonstration of state-of-the-art military systems for battlefield visualization and real-time, man-in-the-loop, networked simulation. David Wray, of DMSO, provided hours of videotaped visual simulations for the committee to examine and excerpt. Several volunteers set up and operated a variety of entertainment and military demonstration systems during the 1996 workshop to provide participants with hands-on experience: Charles Benton of Technology Systems Inc., Michael Bilodeau of Spectrum HoloByte Inc., Steven Carter of Thrustmaster Inc., Leon Dennis of the Armstrong Laboratories at Wright Patterson Air Force Base, Brian Kalita of BBN Corporation, and Greg Lutz of Motorola's Government Electronics Division. Robin Scheer, of Spectrum HoloByte Inc., worked tirelessly to arrange the entertainment demonstrations and to contact participants for the strategy and war games session of the workshop. Fred Zyda orchestrated audiovisual presentations during the workshop, demonstrated video games for participants when called upon, and selected video clips and edited the videotape for the "Introductory Commonalities" presentation.

Finally, thanks are due the sponsors of this study. Anita Jones, as director of defense research and engineering, conceived of the project and ensured its realization. James Hollenbach, Mark Jefferson, and Judith Dahmann of DMSO, with support from Terry Hines, of the MITRE Corporation, provided necessary guidance and support for the project and facilitated the participation of the defense community in its completion.

                    Michael Zyda, *Chair*
                    Committee on Modeling and Simulation:
                    Opportunities for Collaboration Between the

Contents

EXECUTIVE SUMMARY 1

1 INTRODUCTION 13
    Defense Modeling and Simulation, 14
    Modeling and Simulation in the Entertainment Industry, 19
    Connections Between Defense and Entertainment, 23
    Notes, 30

2 SETTING A COMMON RESEARCH AGENDA 32
    Technologies for Immersive Simulated Environments, 32
        Experiential Computing in DOD, 33
        Experiential Computing in the Entertainment Industry, 35
        Research Challenges, 35
    Networked Simulation, 44
        Applications, 44
        Research Challenges, 44
    Standards for Interoperability, 52
        DOD Efforts in Interoperability, 54
        Interoperability in the Entertainment Industry, 58
        Research Areas, 60
    Computer-generated Characters, 64
        Computer-generated Characters in Entertainment, 65
        DOD Applications of Computer-generated Characters, 68
        Common Research Challenges, 69

Tools for Creating Simulated Environments, 73
  Entertainment Applications and Interests, 74
  DOD Applications and Interests, 75
  Research Challenges, 76
Conclusion, 79
Notes, 79

3 SETTING THE PROCESS IN MOTION 84
  Overcoming Cultural Barriers, 84
    Different Business Models, 85
    Facilitating Coordination and Cooperation, 88
  Human Resources, 92
  Maintaining the Research Base, 97
  Concluding Remarks, 99
  Notes, 100

APPENDIXES

A WORKSHOP AGENDA 105

B WORKSHOP PARTICIPANTS 107

C BIOGRAPHICAL SKETCHES OF COMMITTEE MEMBERS 110

D POSITION PAPERS 115

Executive Summary

Modeling and simulation technology has become increasingly important to both the entertainment industry and the U.S. Department of Defense (DOD). In the entertainment industry, such technology lies at the heart of video games, theme park attractions and entertainment centers, and special effects for film production. For DOD, modeling and simulation technology provides a low-cost means of conducting joint training exercises, evaluating new doctrine and tactics, and studying the effectiveness of new weapons systems. Both the entertainment industry and DOD are aggressively pursuing development of distributed simulation systems that can support Internet-based games and large-scale training exercises. These common interests suggest that the entertainment industry and DOD may be able to more efficiently achieve their individual goals by working together to advance the technology base for modeling and simulation. Such cooperation could take many forms, including collaborative research and development projects, sharing research results, or coordinating ongoing research programs to avoid unnecessary duplication of effort.

This report summarizes the results of a workshop, convened by the National Research Council's Computer Science and Telecommunications Board, that brought together members of the entertainment and defense industries to discuss research interests in modeling and simulation. The workshop revealed several areas in which the entertainment industry and DOD have common interests (see Box ES.1). This report examines the research challenges in these areas with an eye toward identifying

BOX ES.1
Research Areas of Interest to the Entertainment Industry and the Defense Modeling and Simulation Community

Technologies for Immersion
- *Image generation*—graphics computers capable of generating complex visual images.
- *Tracking*—technologies for monitoring the head position and orientation of participants in virtual environments.
- *Perambulation*—technologies that allow participants to walk through virtual environments while experiencing hills, bumps, obstructions, etc.
- *Virtual presence*—technologies for providing a wide range of sensory stimuli: visual, auditory, olfactory, vibrotactile, and electrotactile.

Networked Simulation
- *Higher-bandwidth networks*—to allow faster communication of greater amounts of information among participants.
- *Multicast and area-of-interest managers*—to facilitate many-to-many communications while using limited bandwidth.
- *Latency reduction*—techniques for reducing true or perceived delays in distributed simulations.

Standards for Interoperability
- *Virtual reality transfer protocol*—to facilitate large-scale networking of distributed virtual environments.
- *Architectures for interoperability*—network and software architectures to allow scalability of distributed simulations without degrading performance.
- *Interoperability standards*—protocols that allow simulators to work together effectively and facilitate the construction of large simulations from existing subsystems.

Computer-generated Characters
- *Adaptability*—development of computer-generated characters that can modify their behavior automatically over time.
- *Individual behaviors*—computer-generated characters that accurately portray the actions and responses of individual participants in a simulation rather than those of aggregated entities such as tank crews or platoons.
- *Human representations*—authentic avatars that look, move, and speak like humans.

continues

> **BOX ES.1**
> **continued**
>
> - *Aggregation/disaggregation*—the capability to aggregate smaller units into larger ones and to disaggregate them back into smaller ones without sacrificing the fidelity of a simulation or frustrating attempts at interoperability.
> - *Spectator roles*—ways of allowing observers to watch a simulation.
>
> **Tools for Creating Simulated Environments**
> - *Database generation and manipulation*—tools for managing and storing information in large databases, to allow rapid retrieval of information, feature extraction, creation, and simplification.
> - *Compositing*—hardware and software packages that allow designers to combine images taken from different sources (whether live-action footage or three-dimensional models) and to facilitate the addition or modification of lighting and environmental effects.
> - *Interactive tools*—hardware and software systems that allow designers to use a variety of input devices (more than mouse and keyboard) to construct models and simulations.

areas for possible cooperation. The report does not attempt to provide answers to existing research questions, nor does it necessarily recommend that cooperative efforts be initiated in the areas discussed. Such decisions need to be made on a case-by-case basis by the individual organizations that might participate.

Cooperative endeavors between DOD and the entertainment industry will face many obstacles. As the workshop revealed, numerous cultural barriers divide the entertainment industry and DOD, and few mechanisms exist to facilitate the sharing of information. If such obstacles can be overcome, these differences could be a source of strength, ensuring a complementarity of interests, capabilities, and approaches that might benefit both communities. Already, the U.S. Marine Corps is evaluating commercial computer games for training purposes, the Army is considering use of game machines as personal training units, and members of the Air National Guard are evaluating the use of commercial flight simulator programs to supplement standard training regimens. Such initiatives suggest that cooperation is possible but only begin to hint at the kinds of benefits that might be achieved through greater collaboration in research, which is the main subject of this report.

Other tasks will also need to be undertaken jointly in order to ensure the future strength of both communities. DOD and the entertainment industry will need to foster the establishment and expansion of education programs to train students in the technical and nontechnical underpinnings of modeling, simulation, and virtual environments. They also will need to ensure the viability of the university research base, which not only will produce these students but will also generate many of the technical advances upon which future entertainment and defense systems will be built.

TOWARD A RESEARCH AGENDA

Workshop discussions revealed several research areas that are of interest to both the entertainment industry and the defense modeling and simulation community: technologies for immersion, networked simulation, interoperability, computer-generated characters, and hardware and software tools for creating synthetic environments. Each of these areas demonstrates sufficient overlap in interest by DOD and the entertainment industry to suggest that common work may be possible, although additional study may be required to fully detail the scale and scope of such work. While both DOD and the entertainment industry have similar interests at the research and technology levels, the applications and end products into which research results will be incorporated may differ in fundamental ways, reinforcing the notion that the most effective forms of cooperation will derive more from early stages of research than from sharing products. Emphasizing cooperation in research over sharing of products also helps avoid many of the concerns about intellectual property and proprietary interests that could impede collaboration between the entertainment industry and DOD.

Technologies for Immersion

Both the entertainment industry and DOD are interested in developing immersive systems that allow participants (whether game players or soldiers) to enter and navigate simulated environments. For DOD such systems can be used to train groups of combatants or, increasingly, individual combatants for particular missions when access to the actual location is either hazardous or just not possible. They can also be used to create virtual prototypes of military systems that designers can walk through and visualize. For the entertainment industry such systems are the basis for virtual reality (VR) experiences being incorporated into location-based entertainment centers, theme parks, and video game centers. Immersive technologies are also finding their way into home applications, prompted by the greater availability of three-dimensional (3D) graphics on personal computers and reductions in the cost of peripheral devices, such as joysticks and head-mounted displays.

Immersive environments benefit from a wide range of technologies that provide the sensory cues necessary for participants to perceive their environments. The most basic of these are image generators that create 3D visual displays of the environment itself. Other technologies include locomotion platforms and unobtrusive bodysuits that allow participants to walk through virtual environments and track their movements and interactions. Such bodysuits are also used to build keyframes for animated characters in film and video game productions.[1] DOD is pursuing these applications as part of its dismounted infantry program, and the Defense Modeling and Simulation Office recently funded relevant work at the U.S. Army's Simulation, Training, and Instrumentation Command. It is also funding work on generating other sensory stimuli in virtual environments: auditory, olfactory, and tactile. Such work is also the focus of the commercial VR industry and parts of the entertainment industry.

The entertainment industry and DOD may also be able to benefit from their complementary approaches to *selective fidelity*. Both communities have learned how to boost the fidelity (or accuracy) of some parts of a simulated environment and limit the fidelity of others while creating an engaging simulation experience. Whereas DOD has tended to emphasize the fidelity of interactions between objects in a simulated environment (using science-based models), the entertainment industry has tended to promote visual fidelity and uses principles of good storytelling to help participants suspend their disbelief about the reality of a synthetic experience (whether a VR attraction or a film). Additional work in these areas, and sharing of approaches, may allow both communities to create more engaging simulated experiences while minimizing the technical demands placed on the system itself.

Networked Simulation

The entertainment industry and DOD face similar challenges in creating networking technologies capable of supporting distributed simulations. DOD has already demonstrated the capability to link thousands of participants into a single training exercise and is working on systems that would engage tens of thousands of participants. Internet-based game companies have recently begun to move fast-action video games onto the Internet and are looking for ways to increase the number of simultaneous players from 10 or 20 to hundreds or perhaps thousands. As the number of participants in Internet-based games and the military's joint training exercises grows, improvements to simulation networks will be needed to ensure that communications between participants can be transmitted in a reliable and timely fashion.

Several technologies can support such requirements. First, DOD and the entertainment industry can pursue ways of expanding the bandwidth of simulation networks to allow more information to be transmitted more quickly. Many of the technologies for doing so will derive from technical advances made by the communications and networking industries, on which both DOD and the entertainment industry rely. Other approaches also are being pursued by DOD and the Internet community, including multicast and software-based area-of-interest managers. These techniques can be used to minimize message traffic across the network by directing copies of a single message to only those recipients who have an interest in seeing it. DOD and the games industry are also developing ways of compensating for the latency of distributed networks through algorithms for predicting the future location and state of other objects and for synchronizing events among different participants.

Standards for Interoperability

Both DOD and the entertainment industry are developing architectures and protocols for linking distributed simulation systems. DOD has promoted the development of standards for distributed interactive simulation (DIS) that specify the protocols such systems should follow. It has also developed a high-level architecture for military simulation systems that allows different simulator platforms to interoperate and enables reuse of existing simulation programs. Commercial industry, in contrast, has developed standards for enabling different types of computer systems to play the same game. Rather than adopt a common set of protocols for allowing games to work with one another, game companies have each tended to develop their own proprietary protocols that allow copies of their own games to play against each other but do not allow them to work with another company's games. Some of these protocols derive from DIS standards but are modified to boost the performance of a particular game. It is not clear that DOD and the entertainment industry will adopt common standards on a wide scale as long as proprietary interests continue to dominate protocol decisions.

Nevertheless, both DOD and the entertainment industry will need to solve common problems in developing their network architectures and protocols; common research into interoperability standards might be beneficial. A careful and considered joint research program is needed that studies the issues involved in designing a common scalable network software architecture capable of supporting large numbers of players across wide-area networks.[2] Components of this research effort include Web-based interoperability standards that would allow the linking of distributed virtual environments, architectures for "plug-and-play" interoperability that allow different simulation systems to interoperate, and network software architectures for maintaining *persistent universes*—simulated worlds that continue to exist and evolve even after an individual participant leaves the simulation. Little fundamental research is being conducted in these areas by either DOD or the entertainment industry; rather, both communities are concentrating on developing networked simulation systems without addressing the basic issues of the network software architecture.

Computer-generated Characters

The term *computer-generated characters* refers to the broad range of entities in a simulated environment (people, tanks, aircraft, etc.) whose behaviors are controlled wholly or in part by a computer.[3] They include the computerized opponents in computer chess games, digital actors that appear in films and television, and simulated enemy forces in military training exercises. Computer-generated characters are a part of virtually every major DOD simulation and all video games in which players compete against the computer instead of, or in addition to, other players. They attempt to reproduce realistic intelligent human behavior that provides participants with a compelling simulated experience.

Additional research would benefit computer-generated characters in both entertainment and defense applications. Gilman Louie, of Spectrum HoloByte Inc., estimates that games companies allocate up to two-thirds of their development efforts to programming autonomous characters that cannot be reused in other games. DOD, while creating more advanced computer-generated characters, tends to program behaviors into entities, such as tanks crews, that operate according to strict rules of engagement derived from military doctrine. These entities cannot be easily modified or reprogrammed; nor can they accurately portray the behaviors of individual soldiers on a battlefield. Both DOD and the entertainment industry would like to develop computer-generated characters that have adaptable behaviors and can learn from experience. Some research is ongoing, under DOD sponsorship, to apply techniques of artificial intelligence and genetic algorithms to computer-generated characters so that they can achieve these capabilities. Other work is needed to develop adequate models of individual human behaviors and realistic representations of humans in virtual environments. Significant work is needed to develop the capability to define computer-generated characters at a high level of abstraction that will facilitate their reuse.

Tools for Creating Simulated Environments

Workshop participants identified low-cost, easy-to-use hardware and software tools for creating simulated environments as a critical need for entertainment and defense applications of modeling and simulation.[4] Such tools are needed to enable the rapid creation and manipulation of large databases of information describing terrain, buildings, 3D objects, and dynamic features of virtual environments, and to facilitate the compositing of such disparate imagery into a unified simulated world. Although the entertainment industry purchases a wide variety of graphics hardware and software from established vendors, it spends little on improving these tools and instead concentrates on short-term solutions to devise advanced special effects. Many existing systems are expensive and difficult to learn. Additional research is needed to create more interactive tools that allow designers to develop simulated environments using input devices other than keyboard and mouse. In one system described at the workshop, an immersive VR system was developed to allow filmmakers to manipulate computer imagery in real time. Such techniques may have broader applicability in entertainment and defense.

TOWARD GREATER COOPERATION

Promoting cooperation between DOD and the entertainment industry in modeling and simulation will require both communities to overcome cultural barriers that have, to date, isolated them from one another and limited the flows of information between them. Differences in business models will need to be overcome if joint research is to be achieved. At the same time, DOD and the entertainment industry will need to ensure that they take the necessary actions, both individually and jointly, to ensure a continued supply of good people and good ideas for future modeling and simulation efforts. Educational programs are needed to train students in the technical and nontechnical skills that are important to creating effective simulated environments. Fundamental research programs are needed to generate ideas and explore new technologies that are broadly relevant to modeling and simulation. DOD and the entertainment industry will need to solicit additional input from the academic research community to better understand how to accomplish these tasks.

Information Sharing and Technology Transfer

The workshop conducted as part of this study was unique in that it brought together two communities that traditionally have shared little information and transferred little technology between them. Its success attests to the vision of DOD in identifying a potential basis for cooperation and the efforts of the committee to create an atmosphere in which mutually beneficial exchanges of information could occur. For the most part, DOD and the entertainment industry are two different cultures, with different languages and separate communities of researchers and managers. Few opportunities exist for promoting information exchanges between the two communities. The ones that do exist—mostly government efforts to promote commercialization of technologies developed by federal laboratories—have been relatively unsuccessful in creating bridges to the entertainment industry.

Workshop participants suggested that additional mechanisms are necessary for promoting information exchanges on modeling and simulation technology that would benefit both DOD and the entertainment industry—even if they do nothing more than identify research problems that have already been solved. These could take the form of formal collaborative arrangements between entertainment companies and DOD, efforts by individual firms to supply modeling and simulation technology to both communities, or joint research endeavors mediated by a university research center. Experiments are needed to test the viability and effectiveness of these different arrangements. Less formal mechanisms also could be effective. Conferences are the primary mechanism for information exchanges today, but DOD and entertainment industry representatives tend to attend separate conferences. Some progress could be made by encouraging cross-attendance at major conferences within each community or by cross-fertilizing boards of relevant technical and planning groups and establishing a separate symposium to specifically explore topics of interest to both communities. Greater use of the Internet and World Wide Web also might facilitate greater communication.

Human Resources

At the workshop, representatives of the entertainment industry and DOD noted an apparent shortage of talented people with the broad range of skills needed to develop models and simulations. Both communities increasingly have trouble finding programmers with experience in content development and the technical problems associated with multiplayer/multiprocessor games and simulations. Both DOD and the entertainment industry are seeking people who are visually literate: people who are skilled in generating economical, high-quality graphics displays and have a good understanding of human perception so that they can create worlds that have the desired effect on those who experience them.

Additional efforts will be needed to enhance educational programs for visual literacy. At present, only a small number of U.S. universities offer interdisciplinary programs that combine technical and artistic studies. Expansion of such programs will require the creation of interdisciplinary degree programs and reward systems that encourage faculty members to pursue such endeavors. The research community will need to articulate a research agenda that incorporates the perspectives of technical and nontechnical disciplines. Workshop participants believed that DOD and the entertainment industry could use existing funding mechanisms as a means for encouraging the creation of such programs without incurring additional costs.

Preserving the Research Base

Ensuring an adequate supply of new ideas and technologies for future modeling and simulation efforts requires continued support for relevant fundamental research. University research is especially important because it concentrates more heavily on basic than applied research and has the added benefit of educating students, who then disseminate new knowledge throughout the research community and industry when they graduate.

Workshop participants concurred that over the past 20 years the nature of research funding in such fields as computer graphics and networking has changed. University researchers have less freedom to select and pursue research areas they deem interesting. In part because of growing demands for accountability, government-funded projects are often more results-oriented than they used to be, and government agencies are under greater pressure to demand specific project goals and delivery dates for each task. The implication is not only a change in the kind of research many investigators perform, but also a reduction in the quantity of research conducted. Most university researchers reported that they now spend less time on research and more time filling out grant applications and seeking funding. Determining the most suitable method for government support of university research is beyond the scope of this study,[5] but members of the university research community present at the workshop expressed considerable concern about the current trend in government funding.

To date, industry funding has not compensated for changes in federal research funding. Although industry contributions to university research have grown over the past decade, they are still small. Moreover, as in many other industries, entertainment companies tend not to conduct long-term basic research, largely because of short planning horizons and the inability to fully appropriate the results of fundamental research.[6] Entertainment companies tend to obligate most of their research and development expenditures to technical problems related to a particular film or game release rather than to longer-term issues related to future needs. Industry-sponsored research is more closely tied to particular product needs, potentially limiting the scope of inquiry and raising concerns that valuable new information will not be widely disseminated.

Other forms of industry support for university research also appear to be changing. Several workshop participants observed that computer firms do not donate as much equipment to university laboratories as in the past. University representatives noted that they rely heavily on such donations to acquire state-of-the-art equipment for research projects and education. Their inability to attract such donations affects not only the quality of research but also the training of students.

Recent trends in federal and industry funding for university research in modeling and simulation mirror those of other scientific and technical fields. National support for research and development (R&D) is undergoing a period of transformation. Interest in reducing the federal budget deficit and in realigning defense needs to match the challenges of the post-Cold War environment will continue to put pressure on federal funding for R&D. Increased competition seems to be changing the nature and structure of industrial R&D. Such issues must be addressed at the national level to ensure the continued viability of the technology base for modeling and simulation.

NOTES

1. The need for research into lightweight tracking technology is fully described in another National Research Council report. See *Virtual Reality: Scientific and Technological Challenges*, Nathaniel I. Durlach and Anne S. Mavor, eds., National Academy Press, Washington, D.C., 1995.

2. The term *network software architecture* encompasses both network architecture and software architecture to indicate that the problems of network bandwidth and limited processor cycles must be solved together to achieve scalability.

3. The terms *computer-generated forces, autonomous forces, semiautonomous forces*, and *autonomous agents* all refer to computer-generated characters. The first three terms are widely used throughout the defense community; the term *autonomous agents* refers to a broader class of entities used for seeking relevant information on computer networks as well as generating computerized opponents for game players.

4. The need for such tools is also described in *Virtual Reality: Scientific and Technological Challenges*, note 1 above.

5. The National Research Council's Computer Science and Telecommunications Board is conducting two other studies that may more fully investigate this topic. The first will look retrospectively at the role of government, industry, and universities in key innovations in information technology. The second will look prospectively at institutional arrangements for ensuring the continued leadership of the U.S. information technology industry.

6. The difficulty of appropriating profits from investments in basic research has been a long-standing issue in economics and management. For greater discussion of this topic, see Teece, David J., 1988, "Profiting from Technological Innovation: Implications for Integra-

tion, Collaboration, Licensing, and Public Policy," in *Readings in the Management of Innovation,* Michael L. Tushman and William L. Moore, eds., Ballinger Publishing Company, Cambridge, Mass., pp. 621-647; and Levin, Richard et al., 1987, "Appropriating the Returns from Industrial Research and Development," *Brookings Papers on Economic Activity,* No. 3, pp. 783-831.

1

Introduction

From three-dimensional (3D) graphics on home video games to the special effects and animation sequences created for feature films, it is apparent that the entertainment industry has emerged as an innovative source of modeling and simulation technology.[1] The U.S. Department of Defense (DOD) has an even longer history of investing in modeling and simulation to support objectives such as training and analysis programs and has supported the development of many of the fundamental computer graphics and networking technologies that underlie both military and entertainment applications of modeling and simulation. Though the two communities differ widely in their structures, incentives, and motivations, opportunities may exist for the entertainment industry and the defense modeling and simulation community to work together to advance the state of the art in modeling and simulation technology. By sharing research results, coordinating research agendas, and working collaboratively when necessary, the entertainment industry and DOD may be able to more efficiently and effectively build a technology base for modeling and simulation that will improve the nation's security and economic performance.[2]

This report explores the potential for greater cooperation between the entertainment industry and DOD in modeling and simulation. It draws heavily on a workshop convened by the Computer Science and Telecommunications Board of the National Research Council in October 1996 that brought together representatives of the entertainment industry and the defense modeling and simulation community to discuss issues of mutual interest and identify areas for cooperation. The report summarizes major uses of modeling and simulation technology in both defense and entertainment applications, outlines research areas in which the entertainment and defense modeling and simulation communities share a common interest, and identifies other issues that must be addressed to facilitate cooperation and ensure the viability of the technology base for modeling and simulation. It does not explicitly consider the degree to which DOD can adopt commercial off-the-shelf technologies for its own purposes; rather, it examines opportunities for conducting research that could benefit both military and entertainment applications.

As the report demonstrates, the potential exists for greater cooperation between the entertainment industry and DOD, but collaboration may not be easy to achieve. The entertainment industry and DOD have vastly different cultures that reflect different business models, capabilities, and objectives. It is unlikely that the cultures will converge, and bridging them may be difficult. Nevertheless, these differences can be a source of strength. DOD's research efforts and those of the entertainment industry are in many ways complementary rather than contradictory. Whereas DOD's research and development efforts are well funded (by industry standards), meticulously planned, and forward looking, the entertainment industry's efforts are diverse, fast paced, and market oriented. If cultural barriers can be overcome, the resulting cooperation could enable the two communities to leverage each other's strengths to develop a stronger technology base that will allow both to more easily achieve their individual objectives for modeling and simulation.

DEFENSE MODELING AND SIMULATION

DOD uses modeling and simulation for a variety of purposes, such as to train individual soldiers, conduct joint training operations, develop doctrine and tactics, formulate operational plans, assess war-fighting situations, evaluate new or upgraded systems, and analyze alternative force structures (see Box 1.1). The technology also supports the requirements of other critical defense needs such as command, control, and communications; computing and software; electronics; manpower, personnel, and training; and manufacturing technology. As a result of this breadth, defense models and simulations range in size and scope from components of large weapons systems through system-level and engagement-level simulations, to simulations of missions and battles, and theater-level campaigns. DOD's Defense Modeling and Simulation Office (DMSO) coordinates military modeling and simulation programs on an interservice level and has played a key role in developing a standard architecture for military simulations. Each of the military services also has a designated

BOX 1.1
Defense Applications of Modeling and Simulation

DOD's efforts in modeling and simulation generally support three major functions: training, analysis, and acquisition. The vision for each of these areas is outlined below.

- *Training.* Warriors of every rank will use modeling and simulation to challenge their skills at the tactical, operational, or strategic level through the use of realistic synthetic environments for a full range of missions, including peace keeping and the provision of humanitarian aide. Huge exercises, combining forces from all services in carefully planned combined operations, will engage warriors in realistic training without risking injury, environmental damage, or costly equipment damage. Simulation will enable leaders to train at scales not possible in any arena short of full-scale combat operations, using weapons that would be unsafe on conventional live ranges. Simulation will also be used to evaluate the readiness of armed forces. The active duty and reserve components of all forces will be able to operate together in synthetic environments without costly and time-consuming travel to live training grounds.
- *Analysis.* Modeling and simulation will provide DOD with a powerful set of tools to systematically analyze alternative force structures. Analysts and planners will design virtual joint forces, fight an imaginary foe, reconfigure the mix of forces, and fight battles numerous times in order to learn how best to shape future task forces. Not only will simulation shape future force structure, it will be used to evaluate and optimize the course of action in response to events that occur worldwide. Modeling and simulation representations will enable planners to design the most effective logistics pipelines to supply the warriors of the future, whether they are facing conventional combat missions or operations other than war.
- *Acquisition.* Operating in the same virtual environments, virtual prototypes will enable acquisition executives to determine the right mix of system capability and affordability prior to entering production. Fighting synthetic battles repeatedly while inserting new systems or different components will help determine the right investment and modernization strategy for future armed forces. Models and simulations will reduce the time, resources, and risks of the acquisition process and will increase the quality of the systems produced. In addition, modeling and simulation will allow testers to create realistic test and evaluation procedures without the expense and danger of live exercises. "Dry runs" of live operational tests will minimize the risks to people, machines, and testing ranges. Modeling and simulation will enhance information sharing among designers, manufacturers, logisticians, testers, and end users, shortening the system development cycle and improving the integrated product team development process.

SOURCE: Defense Modeling and Simulation Office, position paper prepared for this project; see Appendix D.

office to serve as its single point of contact on all modeling and simulation matters.³

DOD's interest in modeling and simulation is growing. The 1997 Defense Technology Area Plan identifies modeling and simulation as one of five key areas of information technology critical to U.S. defense needs and projects growth in funding for enabling technologies from $169 million in 1998 to $280 million in 2003.⁴ Most of these initiatives will be orchestrated by the DMSO, the Defense Advanced Research Projects Agency (DARPA), the Defense Special Weapons Agency, and the U.S. Army. Several other projects are under way by the Army, Navy, Air Force, and Marines to individually and jointly develop simulation systems. Overall acquisition of training systems by military departments exceeds $1.5 billion per year, including both trainers for specific systems (such as the B-2 bomber) and simulators for the integrated performance of a variety of weapons systems. Development of individual simulation systems can easily cost between $100 million and $1 billion (see Table 1.1).⁵

DOD's growing interest in modeling and simulation derives from several factors. Changes in the geopolitical environment are requiring the military to plan for actions not only in traditional regions of conflict, such as the former Soviet Union and the Middle East, but elsewhere in the world as well. Thus, DOD needs to be able to rapidly model varied locations and scenarios to assist in training troops. In addition, DOD is being asked to conduct a broader range of missions (such as drug interdiction and peacekeeping), to defend against new types of threats, and to coordinate joint operations that cross service and national boundaries. Each of these missions requires the development of new doctrine and tactics as well as training. At the same time, advances in information technology have lowered the cost of computer-based models and simulations, making modeling and simulation a cost-effective alternative to live training. Simulated training exercises do not require the space or transportation needed for a live training exercise, nor do they have the environmental impact of live training exercises.⁶ Already, DOD's modeling and simulation activities, such as SIMNET, have helped the services get away from major field exercises that required the agency to move large numbers of people around. In the future, DOD hopes to use modeling and simulation to provide readily available, operationally valid environments for use by all DOD components. It would like users to have daily access to war-fighting scenarios from their offices, in the same places that they normally work.

DOD has developed a Modeling and Simulation Master Plan as a first step in directing, organizing, and concentrating its modeling and simulation activities. It is intended to be dynamic and flexible, evolving

TABLE 1.1 Large DOD Development Programs in Modeling and Simulation

Project Name	Description	Estimated Program Cost ($ millions)
Close Combat Tactical Trainer	Networked simulation system for training army mechanized infantry and armor units. It is composed of various simulators that replicate combat vehicles, tactical vehicles, and weapons systems interacting in real time with each other and semiautonomous opposing forces.	$ 846
Battle Force Tactical Training	Tactical training system for maintaining and assessing fleet combat proficiency in all warfare areas, including joint operations. It will train at both the single-platform and battle group levels.	165
Warfighter's Simulation 2000	Next-generation battle simulation for training Army commanders and battle staffs at the battalion through theater levels. It has a computer-assisted exercise system that links virtual, live, and constructed environments.	172
Joint Tactical Combat Training System	Joint effort by the Navy and Air Force to create a virtual simulation at the battle group level in which combat participants will interact with live and simulated targets that are detected and displayed by platform sensors.	270
Synthetic Theater of War (STOW) Advanced Concept Technology Demonstration	STOW is a program to construct synthetic environments for numerous defense functions. Its primary objective is to integrate virtual simulation (troops in simulators fighting on a synthetic battlefield), constructive simulation (war games), and live maneuvers to provide a training environment for various levels of exercise. The demonstration program will construct a prototype system to allow the U.S. Atlantic Command to quickly create, execute, and assess realistic joint training exercises.	442
Joint Simulation System (core)	A set of common core representations to allow simulation of actions and interactions of platforms, weapons, sensors, units, command, control, communications, computers, and intelligence systems, etc., within a designated area of operations, as influenced by environment, system capability, and human and organizational behavior.	154
Distributed Interactive Simulation	A virtual environment within which humans may interact through simulation at multiple sites that are networked using compliant architecture, modeling, protocols, standards, and databases.	500
TOTAL		$2,549

SOURCE: U.S. Department of Defense, Office of the Inspector General. 1997. *Requirements Planning for Development, Test, Evaluation, and Impact on Readiness of Training Simulators and Devices*, a draft proposed audit report, Project No. 5AB-0070.00, January 10, Appendix D.

as the technology matures and consensus develops on policy and programmatic issues.

The first objective of the master plan is establishment of a common technical framework to facilitate interoperability among simulations and the reuse of simulation components. The key to this effort is the development of a standard architecture for defense simulations, the High-Level Architecture, with which all defense models and simulations must comply. This architecture was designed to allow DOD to meet its vision of constructing a rapidly configured mix of computer simulations, actual war-fighting systems, and weapons systems simulators geographically distributed and networked, involving tens of thousands of entities to support training, analysis, and acquisition. Such simulations would be used both to train individuals to perform particular tasks, to interpret data, and to make decisions, and to help groups of individuals (tank crews, fighter squadrons) work together as a team.

The second objective of the master plan is to provide timely and authoritative representation of systems (aircraft, ground vehicles, ships, communications systems, etc.), the natural environment (air, space, land, sea, weather, and battle effects), and individual human behaviors.

Efforts are under way to create databases that would allow just-in-time generation of integrated and consistent environmental data to support realistic mission rehearsals anywhere in the world, including locations that are difficult to access or that are operationally dangerous. This work is attempting to develop the capability to generate—with minimal editing—synthetic representations of geographic surfaces that incorporate relevant surface features (trees, rocks, etc.) and to create model-based software tools for feature extraction. Achieving these goals will ensure, for example, that weather fronts that bring rain or snow to an area will affect the transit rate of vehicles and troops and that wind patterns will move trees, create waves, and alter dispersal patterns of smoke and dust. These effects will not only help increase the realism of DOD simulations (and, hence, more realistic training and analysis) but will also allow simulation of different seasonal conditions.

Other objectives include the establishment of a robust infrastructure to meet the needs of simulation developers and end users. The infrastructure will include resource repositories—virtual libraries—and a help desk for users. The goal is to provide common services and tools to simulation developers to further reduce the cost and time required to build high-fidelity representations of real-world systems and processes. Such tools will enable the construction of realistic simulations that interact with actual war-fighting systems to allow combatants to rehearse missions and train as they will fight. It could also facilitate development of virtual prototypes that could be evaluated and perfected with the help of real war fighters before physical realizations are ever constructed. Such virtual prototypes could have applications outside defense, such as in city planning, architecture, and design (see discussion of database generation and manipulation in the "Tools for Creating Simulated Environments" section of Chapter 2).

The final objective of the plan is to share the benefits of modeling and simulation. DOD must educate potential users about the benefits of modeling and simulation. To that end, an extensive study is under way to quantify objective data on the cost-effectiveness and efficiency of modeling and simulation in training, analysis, and acquisition applications throughout DOD. Extensive anecdotal data exist, but no concerted effort has been made to demonstrate the return on investment.

MODELING AND SIMULATION IN THE ENTERTAINMENT INDUSTRY

The entertainment industry consists of a varied mix of companies engaged in a broad range of activities, including film, television, radio, recorded music, publishing, performing arts, home entertainment, and video. Companies in these industries are using digital electronic technology for many applications: (1) to deliver existing products, such as video games and video on demand, and potentially to distribute products to audiences that are not reached today; (2) to create electronic games and other forms of digitized material (such as films that have been converted into electronic games); (3) for direct response sales (i.e., home shopping); (4) for new entertainment products that are still in the process of being invented (such as musical books or interactive stories); (5) for location-based entertainment, such as high-tech theme parks based on visual simulation and other offshoots of the aerospace and electronics industries; and (6) for new methods that enhance the quality or lower the costs of producing products (e.g., computer animation systems or virtual reality systems for set design and lighting).[7]

Of these industries, filmmaking, television, video games (including both computer games that run on standard personal computers and console games such as Nintendo, Sega, and Sony systems), and location-based entertainment centers have been most active in adopting modeling and simulation technology. For the most part, these sectors have operated independently of one another, though some blurring of the boundaries is occurring as film studios attempt to develop games based on their movies. Other linkages also exist. Filmmakers and television producers, for instance, often share techniques, technologies, actors, and even content. Companies that produce games are working hand in hand with network service companies to provide networked video games.

These companies play an important role in the U.S. economy. Sales of video games and consoles, such as the Sony PlayStation, Nintendo 64, and Sega Saturn, were expected to surpass $4.3 billion in 1996.[8] Game boxes themselves accounted for nearly $3.6 billion.[9] Such devices are attractive to many game players because they sell for roughly $200 compared with $2,000 for a typical personal computer (PC). Nintendo expected to sell out its production of 1 million Ultra 64 machines in 1996, and Sega sales also were expected to reach nearly 1 million. Game boxes themselves do not usually generate significant profits, but they pave the way for sales of game cartridges. Sales of game software for personal computers are also rising and were expected to grow 20 percent in 1996 to $1.2 billion. Part of the increase is the rise in on-line game sites. Though Internet-based games were expected to generate only $90 million in 1996, they are projected to generate $1.6 billion by 2000.[10] The film industry generated another $22 billion in revenues. Box office receipts totaled almost $6 billion in 1996,[11] with video tape rentals at $16 billion.[12] These figures do not include revenues from merchandise related to films, such as toys, games, and clothing. Such revenues often exceed box office receipts.

In some areas, modeling and simulation technology has already enabled firms to regain their competitiveness internationally. As Ed Catmull of Pixar Animation Studios noted at the workshop, technology saved the animation industry. Most U.S. animation went overseas in the 1980s as studios looked for ways to cut labor costs. The advent of electronic animation technologies (such as those that made the computer-animated film *Toy Story* possible), however, has allowed U.S. firms to win back animation; foreign competition is seen as less of a threat to the U.S. industry. In fact, U.S. firms are now raiding other countries for talent.

Technology will continue to transform the entertainment industry in myriad ways, many of which will be unpredictable over the long term. Nevertheless, certain trends are already apparent. Video games are moving onto the Internet, creating a new way to play games and driving changes in the games themselves (see Box 1.2 and Choudhury et al., 1997[13]). A handful of companies are putting the infrastructure in place for game companies like id Software, Spectrum HoloByte, Acclaim, and others to move their games out of their constricted single-player mode into a worldwide, networked, real-time, multiplayer domain. The Total Entertainment Network (TEN), for example, allows subscribers to play on-line versions of *Duke Nukem 3D, Quake, Command and Conquer, Warcraft*, and *Deadlock*. More games are added regularly. In its first three months TEN garnered more than 14,000 subscribers who pay $14.95 a month to access its Internet-based games. MPath Interactive, another entrant into the on-line games industry, offers an on-line version of *Quake* and recently agreed with Hasbro Interactive to put versions of classic

BOX 1.2
Next-Generation Video Games

The intent of on-line gaming is to create massively networked games in which hundreds, if not thousands, of players can play on the same virtual world simultaneously. Gilman Louie of Spectrum HoloByte Inc. predicts that the next generation of computer games will be designed around a block of server code that will enable off-the-shelf products to be played either in single-player or multiplayer mode. In multiplayer mode the player will enter a persistent universe that will run 24 hours a day, 365 days a year. Players will be able to join a game whenever they want and in whatever role they want (tank commander, fighter pilot, etc.). They will be immersed in an environment of teammates and adversaries controlled by other players and by the computer, with the distinctions between the two becoming increasingly hard to detect. When players enter the game, they will replace units being controlled by the computer.

Behind the scenes, Louie says, will be a campaign engine that will move all of the computer-generated forces, monitor other players' moves, and serve up new tasks for the players. Most video games currently use a progression of linearly scripted missions to advance a player through the game. Each mission contains predetermined outcomes and paths. Players conduct each mission over and over again until they successfully graduate to the next level of play. The campaign engine allows a different structure, in which story lines and missions are dynamic and outcomes are not predetermined. Each play of the game influences the next. If a player is first assigned a mission to destroy a bridge but fails, the next mission may be to provide support to friendly tanks that are being engaged by an enemy that just crossed the bridge. The campaign engine will aggregate and disaggregate units as players encounter them (e.g., converting an icon for a tank unit into four separate tanks and vice versa). Disaggregated units will be controlled by the computer until the player leaves the area and they are re-aggregated.

Game servers will also support multiple dissimilar products, like different aircraft types and ground-based vehicles, that users will buy at retail. Users will be able to download upgrades to their vehicles, such as new avionics, weapons systems, and better automated individuals or units. Every month new scenarios will be generated to enhance game play. These scenarios could include new terrain or specially scripted missions. The system will be designed so that the visual display system of the games will be totally independent from the server, so that upgrades to the visual systems will be tied to each individual simulator rather than being an inherent part of the networked architecture. The network will be totally object oriented to facilitate seamless upgrades and enhancements.

SOURCE: Gilman Louie, chairman, Spectrum HoloByte Inc., Alameda, Calif.

Parker Brothers board games, such as *Monopoly*, *Scrabble*, and *Risk*, on line. MPath offers subscribers who have multimedia PCs the ability to talk to other players during games. Players speak into microphones attached to their PCs, and the MPath software digitizes their voices and transmits them over the Internet to other players. Players without microphones can communicate by typing messages.

A growing number of companies are entering the market for location-based entertainment, using virtual reality (VR) technologies as the centerpiece of their centers. Location-based centers generally provide a specific entertainment attraction, often accompanied by a cafe, bar, or restaurant. A recent compilation listed 153 VR entertainment centers worldwide, ranging from restaurants or cafes with one or two VR units to larger facilities and theme parks with up to 40.[14] In some of these centers, participants don a head-mounted display and enter a 3D world through which they navigate with a joystick or ski down a simulated mountain. In others, groups of players sit in pod-like facsimiles of military aircraft and fly through a simulated landscape, engaging enemy targets and communicating with a control tower. Further advances in technology combined with reductions in price could enable simple VR technologies to enter homes. Already, companies such as Thrustmaster Inc. are marketing mock-ups of aircraft cockpits for home use in conjunction with flight simulator games designed for PCs.

Industry analysts, such as John Latta of 4th Wave Inc., see CD-ROMs, "Internetworking," and interactive television as the primary means of delivering entertainment in the 1990s, although motion-based simulators, VR experiences, and large-format films also will contend. With the expansion of infrastructure and content, home entertainment is becoming more popular; spending for in-home entertainment far exceeds that for out-of-home entertainment. The emerging market for 3D PC applications may reinforce this trend.

As the price of 3D image generators continues to decline and performance improves, 3D graphics will become a key feature of home PCs. Latta predicts that by the end of 1998, most new PCs will include 3D graphics accelerators. Some 30 to 40 companies are designing or producing 3D video chips for PCs. The market for 3D accelerators is predicted to grow from 5,200 chips in 1994 to 36 million in 1999. Applications will range from home video and PC games to 3D tools, such as animation and modeling software, VR, multimedia 3D, and interactive television. Venture capitalists have already pumped $200 million into 3D start-up firms, and over 25 companies have invested at least $1 million apiece in multiplayer games, but the market must still be created, and developers have little control over development of the infrastructure.[15]

Film companies, too, are pursuing innovations in information technology. Disney's animated feature films have been all digital for several years now and, as demonstrated by *Toy Story* (produced in association with Pixar Animation Studios), are achieving realistic 3D images. A growing number of nonanimated films, such as *Terminator 2: Judgment Day, Jurassic Park,* and *Casper,* incorporate digital effects and characters. Companies such as Boss Film Studios, Digital Domain, and Industrial Light and Magic continue to improve the realism of these effects and are developing ways to digitize real actors for use in stunts and other special effects.[16] In addition, virtually every major Hollywood studio has established a subsidiary to create interactive products, typically computer adventure games based on movies. Record companies, too, seeing the music production innovations being led by specialized multimedia companies, are exploring interactive media. The new Academy of Interactive Arts and Sciences was established to confer awards in the field; the Houston International Film Festival has established new prizes for interactive multimedia products; and the American Film Institute's computer-based graphics, editing, and multimedia classes are overflowing.[17]

CONNECTIONS BETWEEN DEFENSE AND ENTERTAINMENT

The idea of linking research efforts in DOD and the entertainment industry is not as far fetched as it might first appear. Connections between the two communities stretch back over the decades and have taken many forms, from sharing products, to sharing technologies, to sharing people.[18] The entertainment industry now rests on a technological foundation laid by large amounts of government-funded research and infrastructure, including advanced computing systems, computer graphics, and the Internet. In the area of computer graphics, for example, early DOD funding resulted in development of the geometry engine, about 1979 (see Box 1.3). This technology has since been incorporated into a number of game devices, such as the new Nintendo 64 machine. Similarly, early advances in networking in the late 1950s and 1960s laid the groundwork for the ARPANET, which grew into today's Internet and has become the foundation of today's growing networked games industry. As these examples demonstrate, 20 years or more often pass before DOD-sponsored research generates new technology that is incorporated into a new product.[19]

Technology has also flowed back to DOD. The agency has benefited from the entertainment industry's constant attempts to lower the price/performance ratios for image generation, networking technologies, and content development tools, to name a few areas. It has also benefited from new ideas pioneered by the entertainment industry. The first aircraft simulator created by Edwin Link—which became the basis for the

> **BOX 1.3**
> **Defense Funding and the Roots of Computer Graphics**
>
> Defense funding, channeled primarily through the Defense Advanced Research Projects Agency (DARPA) and the Office of Naval Research (ONR), played a key role in creating computer graphics technologies that now lie at the heart of many entertainment and business applications. Programs sponsored by DARPA and the National Science Foundation supported research activities at the University of Utah, North Carolina State University, Ohio State University, California Institute of Technology, and Cornell University. In the early 1970s, researchers at the University of Utah developed techniques for creating three-dimensional (3D) images that were more realistic than wire frame images drawn with lines. Work by G.S. Watkins and others resulted in a faster way of determining which parts of objects were hidden and drawing only those visible from the viewer's vantage point. Work by Henri Gouraud, Bui-Tuong Phong, and others resulted in techniques for smoothly shading curved surfaces. Additional work at the New York Institute of Technology created the basis for software used to render graphics images. Industry still uses the basic algorithms developed at the University of Utah for simple light calculations, in both software and commodity graphics hardware. More sophisticated rendering packages that exploit algorithms developed at universities are used in the film and animation industries, as well as in flight simulators and computer-aided design. These include the Renderman system used by Pixar Animation Studios for such animated films as *Toy Story*.

military's flight simulator program—was originally sold to amusement parks as an entertainment device (see position paper by Jacquelyn Ford Morie in Appendix D). Game machines are now being considered for military training.[20] Discussed below are several projects currently under way or under consideration to modify commercial hardware and software for military training applications:

• Peter Bonanni, of the Virginia Air National Guard, for example, has been working with Spectrum HoloByte Inc. to modify the *Falcon 4.0* flight simulator game for military training. Budgetary pressures and worldwide deployments have caused some segments of the armed forces to face true training shortfalls for the first time in decades. U.S. Air Force active duty and reserve squadrons, for example, have experienced a reduction in training sorties of up to 25 percent as a direct result of deployments in support of contingency operations over Iraq and Bosnia (see position paper by Peter Bonanni in Appendix D). Since conducting realistic training is impossible on most of these missions, simulators provide

> The first implementations of virtual reality in 1968 also derived from government funding from ONR, the Air Force, and the Central Intelligence Agency, with contributions from Bell Labs. With such support, Ivan Sutherland, then at Harvard University, developed the head-mounted display as well as stereo and see-through displays, head tracking, and a handheld 3D cursor. Such devices are now used in video games and in rapid prototyping systems for design, architecture, and scientific visualization.
>
> The hardware used in computer graphics also traces its roots to federal funding. While a graduate student at Utah, Jim Clark and his adviser, Ivan Sutherland, pursued research in 3D graphics hardware with government funding. After joining the faculty at Stanford University, Clark received support from DARPA's Very Large Scale Integrated Circuit Program for the Geometry Engine Project, which developed techniques for producing custom integrated circuits for cost-effective high-performance graphics systems. The resulting technology formed the basis of Silicon Graphics Inc., which has become a leading supplier of graphics computers to the defense and entertainment industries.
>
> ---
>
> SOURCE: Computer Science and Telecommunications Board, National Research Council. 1995. *Evolving the High-Performance Computing and Communications Initiative to Support the Nation's Information Infrastructure,* National Academy Press, Washington, D.C., pp. 20-21; McCracken, Edward R. 1997. "Computer Graphics: Ideas and People from America's Universities Fuel a Multibillion-Dollar Industry," in *Computing Research: A National Investment for Leadership in the 21st Century.* Computing Research Associates, Washington, D.C., pp. 11-15.

the only realistic training alternative. Unfortunately, most of the simulators in use today are very expensive, are limited to single-crew training, and are not deployable. As a result, pilots have few opportunities for training while on deployment, and proficiency declines as the deployment wears on. The problem also is occurring in other military services as the trend to use U.S. forces in peace-keeping roles accelerates. Low-cost commercial simulators may be a near-term solution to this military training problem. Though lacking the fidelity to allow fighter pilots to practice certain skills, such as properly timing the release of a weapon to ensure the greatest probability of intercept, low-cost simulators may allow pilots to maintain familiarity with the layout of cockpit and throttle controls and to "keep their heads in the game." According to Peter Bonanni, games such as *Falcon 4.0* realistically mimic the look and feel of real military aircraft and allow users to play against computer-generated forces or, in a networked fashion, against other pilots, which facilitates team training opportunities.

- The U.S. Marine Corps has initiated a program to evaluate commercial war games software for use in training. The Marine Corps sees such games as a low-cost way of engaging soldiers in daily decision-making exercises to help improve their tactical decision-making capabilities. The Corps' Combat and Development Command in Quantico, Virginia, evaluated close to 30 games in 1995 for their potential teaching value, examining their cost as well as technical issues such as memory and processor requirements, data accuracy, and ease of use; multiplayer capabilities, level of game play (strategic, operational, or tactical); relevance to the Marine Corps Task Map; and compatibility with Marine Corps doctrine and tactics. In future tests an education specialist will evaluate the educational merits of each game and determine whether it produces negative training. The evaluations to date have found that while no war game was capable of producing a "robust simulated combat environment," several offered potential for training: *Harpoon2, Tigers on the Prowl, Operation Crusader, Patriot,* and *DOOM*.[21] The Computer War Game Assessment Group recommended the use of these games, and the Marine Corps commandant has authorized commanders to permit these games to be loaded onto government computers and to allow Marines to play them during duty hours.[22] The Marine Corps has already begun using *DOOM* for training four-person fire teams. Users play in a networked environment that allows them to cooperate, listen, and make decisions quickly.[23] The game has been modified from its original version to include fighting holes, bunkers, tactical wire, "the fog of war," and friendly fire, as well as Marine Corps weapons, such as the M16(a1) rifle, M-249 squad automatic weapon, and M-67 fragmentation grenades. Such activities are not viewed as a replacement for field training but are used in the hope of making field training more efficient.

- The Army Battle Command Battle Laboratory is discussing the possibility of adapting the Nintendo 64 game machine as a low-cost individual training device. The system, which would be developed by Silicon Graphics Inc. and Paradigm Simulation Inc., would represent an alternative to PCs and CD-ROMs. Initial analyses indicate that the Nintendo 64 is less expensive than alternative trainers and offers more interactivity and visual realism. Unit commanders would be able to purchase them in larger quantities than other systems, allowing more soldiers access.[24]

- The Marine Corps has awarded a contract to MäK Technologies to design a video game that can be used for military training as well as home entertainment. The company will use the same game engine in both the military and civilian versions. The military version will add more accurate details about tactics and weapons, while the civilian game will be less demanding. Both versions will allow multiple players to compete against each other over a local-area network or the Internet.[25]

Additional opportunities may exist for DOD and the entertainment industry to share the data and resources used to create simulations. For example, DOD has created a simulation of one of the major tank battles of the Persian Gulf War, *73 Easting*. Part of the effort in creating the simulation was collecting geographic data and information regarding the position of military units and terrain features. There are enough similar data readily available to produce comparable studies of Austerlitz, Waterloo, Gettysburg, Antietam, and other battles. Available in formats that permit the viewer to traverse these battlefields in time as well as space, these databases could become staples of history courses, officer training programs, and the countless clubs and societies that cherish military history or stage reenactments. By allowing participants to alter the course of the battles, such simulations could be even more attractive to DOD and the public at large.

To date, the flows of technology between the defense and entertainment industries have been largely uncoordinated. Many derive from large investments the government made in fundamental research and infrastructure for its own purposes but that then became the foundations on which entrepreneurs have created whole new industries. The question that must now be asked is whether there is a way to take advantage of future overlap in interest in a more proactive way to encourage the types of interplay that have occurred in the past.

Military and entertainment simulations have markedly different objectives. In entertainment the driving factor is excitement and fun. Users must want to spend their money to use it again and again (either at home or at an entertainment center) and hopefully are willing to tell others about it. Unrealistically dangerous situations, exaggerated hazardous environments, and multiple lives and heroics are acceptable, even desirable, to increase excitement. Defense simulations, on the other hand, overwhelmingly stress realistic environments and engagement situations. The interactions are serious in nature, can crucially depend on terrain features or other environmental phenomena, and generally rely on the user's ability to coordinate actions with other players.

Nevertheless, many of the future challenges that face the movie industry, games industry, and DOD are similar. A striking example of this is multiplayer simulations using real-time 3D graphics. The DOD is interested in this capability for large-scale training exercises; the games industry is interested in networked games that would allow hundreds or thousands of players to participate. The underlying technologies to support these objectives address similar requirements: networking, low-cost graphics hardware, human modeling, and computer-generated characters. Given future trends in defense modeling and simulation and in the entertainment industry, such overlap is likely to occur more frequently in the future.

These similarities suggest that potential exists for DOD and the entertainment industry to leverage each other's modeling and simulation efforts, provided they understanding the fundamental differences and objectives. Simulation, VR, video games, and film share the common objective of creating a believable artificial world for participants. In this context, believability is less a factor of specific content of the environment than of the perception that a world exists into which participants can port themselves and undertake some actions. In film this process is vicarious; in simulation, VR, and gaming it tends to be active, even allowing participants to choose the form for porting themselves into the environment, whether as occupants of a vehicle moving through the environment, as a separate controllable entity, or as a fully immersed human. Their representation can assume whatever form is appropriate for the environment.

Designing and building such worlds require a common set of enabling technologies, regardless of the application (defense or entertainment) to which the worlds will be put. Because they are fundamental to virtually all simulations, these technologies may represent areas in which DOD and the entertainment industry could collaborate on research and early stages of development:

- *Tools for fabricating synthetic environments.* Computer-based tools are needed to efficiently create 3D virtual worlds that can be sensed in multiple ways (visual, auditory, tactile, motion, infrared, radar, etc.). Cost rises as the size space, resolution, detail, and dynamic features (objects that can interact with participants, like doors that can open or buildings that can be razed) of the simulated environment increase. Tools for efficiently constructing large complex environments are generally lacking; existing toolsets are quirky and primitive, require substantial training to master, and often prohibit the environment architect from including all of the attributes desired.
- *Interfaces.* Interfaces provide the portal through which participants interact with a system. They include displays, entry devices such as keyboards or touch-sensitive screens, VR systems, and a host of other input/output devices that link the participant to the simulator. The increase in the richness of the participant's ability to interact with the synthetic environment and other people and agents similarly ported there is especially important as large-scale simulations are constructed.
- *Networking technologies.* Networking technologies enable large numbers of participants to join in a simulation regardless of their physical locations. The network must be able to accommodate the volume of

messages between and among participants in a timely fashion with a minimum amount of delay or latency. These factors are influenced by both the architecture of the network and the protocols for transmitting information. Protocols are needed to minimize message traffic across the network, and service providers need to figure out how they can provide guaranteed levels of service that distinguish between time-critical interactions and lower-priority messages that are less sensitive to time delays.

- *Computer-generated forces and autonomous agents.* Computer-generated forces and autonomous agents control the actions of elements not directly under the control of a human participant in a simulation. They can be adversaries (as in a computer chess game) or companions (controlling a wingman in a flight simulator) and can represent individual players or aggregated forces (such as an enemy infantry division). Computer-generated forces are critical in any simulation intended to be used by an individual participant or in large networked simulations in which it may not always be possible to ensure enough players to control all the necessary entities.[26] Such forces typically strive to display behaviors characteristic of intelligent human participants.

Both DOD and the entertainment industry could benefit from greater collaboration in the above technical areas. The primary benefit of such collaboration would be the development of a technology base that could support modeling and simulation efforts in either defense or entertainment, eliminating redundancies and sharing technical advances. Collaboration could improve the competitive advantage of entertainment companies and the ability of DOD to meet its national security objectives more efficiently than if the two communities continued to operate independently. As Ed Catmull, of Pixar Animation Studios, and Eric Haseltine, of Walt Disney Imagineering, noted, funding from defense agencies such as DARPA had a significant effect on the development of fundamental technologies critical to defense and entertainment; moreover, it helped develop the human resources required to research, develop, and advance those technologies. More formal collaboration may give DOD and the entertainment industry more opportunities to gain greater leverage from each other's research investments to further their own objectives. Such collaboration will become especially important given continued constraints on defense research and development (R&D) spending. Between 1987 and 1996, real DOD expenditures for R&D declined 27 percent, though the largest cuts were allocated to the development portion of the budget; expenditures on basic and applied research remained almost level. Given current attempts to balance the federal budget and realign federal expenditures on defense R&D to reflect a new set of national priorities, it is unlikely that defense R&D budgets will rise significantly in the near future.

Achieving these benefits will require efforts in two areas. First, DOD and the entertainment industry must identify research areas in which they have a common interest. This is something of an exercise in which the two communities plot their research agendas, identify areas of commonality, and determine ways in which the capabilities of each community can best be leveraged to move the field forward. Second, they must find ways to facilitate collaboration between the two research communities. Differences in culture and business practices must be overcome, and mechanisms must be put in place to facilitate information sharing and, perhaps, collaborative research projects. Unless these types of obstacles are overcome, even the best intentions will not produce fruitful results.

NOTES

1. DOD defines a *model* as a physical, mathematical, or otherwise logical representation of a system, entity, phenomenon, or process. It defines *simulation* as a method for implementing a model over time. See U.S. Department of Defense. 1994. "DOD Modeling and Simulation (M&S) Management," Directive Number 5000.59, January 4.

2. Other communities, such as manufacturing, medicine, and education, also might benefit from greater collaboration with the defense and entertainment industries in advancing modeling and simulation technology. This report addresses the possibility of greater cooperation between the entertainment industry and the defense modeling and simulation community only.

3. U.S. Department of Defense, Directive 5000.59, note 1 above.

4. U.S. Department of Defense, Office of the Director of Defense Research and Engineering. 1997. *Defense Technology Area Plan.* DOD, Washington, D.C., May.

5. U.S. Department of Defense, Office of the Inspector General. 1997. "Requirements Planning for Development, Test, Evaluation, and Impact on Readiness of Training Simulators and Devices," a draft proposed audit report. Project No. 5AB-0070.00, DOD, January 10.

6. Concerns over the environmental impact of the annual Return of Forces to Germany (REFORGER) exercise is one of the considerations that has led to a greater reliance on simulation for that program.

7. Computer Science and Telecommunications Board, National Research Council. 1995. *Keeping the U.S. Computer and Communications Industry Competitive: Convergence of Computing, Communications, and Entertainment.* National Academy Press, Washington, D.C., p. 30.

8. Miao, Walter, Access Media International, as cited in Eng, Paul M. 1996. "I Can't Wait to Go On-line and Blow Something Up," *Business Week*, December 23, pp. 70-71.

9. NPD Group, Port Washington, N.Y., as cited in *Business Week,* note 8 above.

10. Jupiter communications, as cited in *Business Week,* note 8 above.

11. Motion Picture Association of America. 1996. *U.S. Economic Review.* Motion Picture Association of America, Washington, D.C.; available on-line at http://www.mpaa.org/htm#_Hlk388150685.

12. Video Software Dealers Association. 1996. *VSDA White Paper—A Special Report on the Home Video Industry.* VSDA, Encino, Calif.; available on-line at http://206.71.226.123/whitepaper/whitpapr.htm.

13. Choudhury, Seema, et al. 1997. *Entertainment & Technology Strategies.* Forrester Research, Cambridge, Mass., April 1.

14. Atlantis Cyberspace, *VR Entertainment Centers*, downloaded February 17, 1996, from www.vr-atlantis.com/lbe_guide/lbe_list2.html.

15. Latta, John. 1996. *DOD & Entertainment: Where Is the Social Experience?* 4th Wave Inc., Alexandria, Va.

16. Parisi, Paula. 1995. "The New Hollywood: Silicon Stars," *Wired*, December, p. 142.

17. Computer Science and Telecommunications Board, *Keeping the U.S. Computer and Communications Industry Competitive*, p. 31, note 7 above.

18. The two communities also have common physical space. Not only are Southern California and Central Florida focal points for both DOD and entertainment industry efforts in modeling and simulation, but the Walt Disney company also announced in August 1996 that one of its divisions would take occupancy of a 200,000-square-foot facility formerly occupied by the Skunk Works division of Lockheed Martin Corp., a high-security division that designed and engineered some of the nation's most guarded defense projects, including the U-2 spy plane. See Newman, Morris. 1996. "A Unit of Disney Finds an Ideal Space Among the Remnants of the Military-Industrial Complex," *New York Times*, August 28, p. D17.

19. Computer Science and Telecommunications Board, National Research Council. 1995. *Evolving the High Performance Computing and Communications Infrastructure to Support the Nation's Information Infrastructure.* National Academy Press, Washington, D.C.

20. Geddes, John, Silicon Valley Science and Technology Office, U.S. Army Research Laboratory, personal communication, November 20, 1996.

21. Marine Corps Modeling and Simulation Management Office, "Computer Based Wargames Catalog," available on-line at http://138.156.4.23/catalog.

22. Marine Corps Modeling and Simulation Management Office, "Computer Based Wargames Catalog," p. 3, note 21 above.

23. Sikorovsky, Elizabeth. 1996. "Training Spells Doom for Marines," *Federal Computer Week*, July 15; available on-line at http://www.fcw.com/pubs/fcw/0715/guide.htm. See also Ackerman, Robert. 1996. "Commercial War Game Sets Spell Doom for Adversaries," *Signal*, July; available on-line at http://www.us.net/signal/archive/July96/commercial-july.html.

24. Geddes, John, Silicon Valley Science and Technology Office, U.S. Army Research Laboratory, personal communication, November 20, 1996.

25. Bray, Hiawatha. 1997. "Battle for Military Video Game Niche On," *Boston Globe*, April 16, p. 1.

26. DOD's experimentation with distributed interactive simulations during the late 1980s resulted in constant pressure to increase the number of participants in simulated exercises. Because there were not enough simulators or participants to populate a typical battlefield scenario (nor did the technical capability exist to efficiently network together large numbers of participants), DOD began to rely on the use of computer-generated forces.

2

Setting a Common Research Agenda

The entertainment industry and the U.S. Department of Defense (DOD) are both interested in a number of research areas relevant to modeling and simulation technology. Technologies such as those for immersive simulated environments, networked simulation, standards for interoperability, computer-generated characters, and tools for creating simulated environments are used in both entertainment and defense applications. Each of these areas presents a number of research challenges that members of the entertainment and defense research communities will need to address over the next several years. Some of these areas may be amenable to collaborative or complementary efforts.

This chapter discusses some of the broad technical areas that the defense and entertainment research communities might begin to explore more fully to improve the scientific and technological base for modeling and simulation. Its purpose is not to provide answers to the research questions posed in these areas but to help elucidate the types of problems the entertainment industry and DOD will address in the coming years.

TECHNOLOGIES FOR IMMERSIVE SIMULATED ENVIRONMENTS[1]

Immersive simulated environments are central to the goals and needs of both the DOD and the entertainment industry. Such environments use a variety of virtual reality (VR) technologies to enable users to directly interact with modeling and simulation systems in an experiential fashion, sensing a range of visual, auditory, and tactile cues and manipulating objects directly with their hands or voice. Such experiential computing systems are best described as a process of using a computer or interacting with a network of computers through a user interface that is experiential rather than cognitive. If a user has to think about the user interface, it is already in the way. Traditional military training systems are experiential computing systems applied to a training problem.

VR technologies can allow people to directly perform tasks and experiments much as they would in the real world. As Jack Thorpe of SAIC pointed out at the workshop, people often learn more by doing and understand more by experiencing than by simple nonparticipatory viewing or hearing information. This is why VR is so appealing to user interface researchers: it provides experience without forcing users to travel through time or space, face physical risks, or violate the laws of physics or rules of engagement. Unfortunately, creating effective experiences with virtual environments is difficult and often expensive. It requires advanced image generators and displays, trackers, input devices, and software.

Experiential Computing in DOD

The most prominent use of experiential computing technology in DOD is in the area of personnel training systems for aircraft and ground vehicles. DOD also has a series of initiatives under way to develop advanced training systems for dismounted infantry that rely on experiential computing. Such programs are gaining increased attention in DOD and will become a primary driver behind the military's efforts to develop and deploy technologies for immersion in synthetic environments. They are being undertaken in coordination with attempts to develop computing, communications, and sensor systems to provide individual soldiers with relevant intelligence information.[2] Experiential computing, as applied to flight and tank simulation, is a mature science at DOD. There are a number of organizations that have extensive historical reference information they can draw on in specifying the requirements for new immersive training systems. These organizations include the U.S. Army's Simulation, Training, and Instrumentation Command (STRICOM) and the Naval Air Warfare Center's Training Systems Division. Experiential computing is something that has been essential to military training organizations for decades.

For traditional training and mission rehearsal functions, the current need is to reduce the cost of immersive systems. Existing mission rehearsal systems based on image generators like the Evans and Sutherland ESIG-4000 serve the Army's Special Operations Forces well, allowing them to fly at low altitudes above high-resolution geo-specific terrain for hundreds of miles and enabling them to identify specific landmarks along their planned flight path to guide them on their actual mission. Unfortunately, these dome-oriented trainers used to cost upward of $30 million, making it impractical to either procure many simulators or to train many pilots. Cost reductions would allow more widespread deployment of such systems.

Experiential computing technologies are being used by the U.S. Navy in both training and enhanced visualization. For battleships an advanced battle damage reporting system allows a seaman in the battle bridge to navigate a three-dimensional (3D) model of his ship to identify where damage has occurred and both where the best escape routes would be for trapped seamen and which routes the rescue and repair crews should take. In another Navy application developed at the Naval Command Control and Ocean Surveillance Center's (NCCOSC's) Research, Development, Test, and Evaluation Division (which is referred to as NRaD), submarines are fitted with an immersive system that generates a view of the outside world for the commander when they are submerged. Since submarine crews cannot normally look outside the boat except when it is on the surface, a virtual window outside provides not only a view of the seafloor (created through the use of digital bathymetric data) but of the tactical environment as well, with other ships, submarines, sonobuoys, and sea life represented clearly and spatially for the commander to gain a better understanding of the tactical and navigational situation.

In the nonimmersive domain, experiential computing technology is being leveraged by both the Naval Research Lab (NRL) and the Army Research Lab (ARL) in the form of a stereoscopic table-based display. This display is known at NRL as the Responsive Workbench and at ARL as the Virtual Sandtable. The Responsive Workbench was invented in 1992 at the German National Computer Science and Mathematics Institute outside Bonn. NRL duplicated the bench and started exploring how it could be used in a variety of applications. The concept of the workbench is simple. The bench itself is a table 6 feet long, 4 feet wide, and standing 4 feet off the floor. The tabletop is translucent, and a mirror sits underneath at a 45 degree angle. A projector behind the table shines on the mirror and up onto the table surface from below, creating a stereoscopic image on the tabletop. Users wear stereoscopic glasses and a head tracker. As they move their heads, the image changes to reflect that motion and objects appear to be sitting, like a physical model, on the table.

An Army application of this technology is a re-creation of the traditional sand table in which forces are laid out and move around to plan strategies and tactics or to review a training exercise. Coryphaeus Software of Los Gatos, California, is commercializing a similar product, the Advanced Tactical Visualization System, which operates with the commercial version of the Responsive Workbench, the Immersive Workbench by Fakespace Inc. Since commanders are used to working with scale models of battlefields and maps, they can easily accommodate this type of display.

Experiential Computing in the Entertainment Industry

The problem with creating effective experiential computing systems is that they demand real-time graphics. In the entertainment industry, return on investment must be considered. The high cost of immersive technologies has slowed their expansion into entertainment settings. Nevertheless, an increasing number of location-based entertainment attractions and home systems are emerging. The majority of the systems in operation fall into one of three categories: (1) arcade systems, (2) location-based entertainment centers, and (3) VR attractions at theme parks. Location-based entertainment centers and arcades boast both stand-alone systems that allow participants to drive down a race course, ski down a mountain, or play virtual golf. Others have networked together flight simulators that allow players to interactively fly through a virtual environment and engage targets (including each other). Disney has developed a VR attraction based on its film *Aladdin*, and Universal Studios has developed a ride based on *Back to the Future*.

Now that the costs of real-time graphics systems are dropping, it is likely that the list of VR experiences for entertainment will expand and that home applications will become more prevalent. Three-dimensional graphics are becoming more widely available on home computers, and the number and variety of peripheral devices, such as throttle-like joysticks and mock-ups of fighter cockpits, are expanding. Continued reductions in cost coupled with increases in capability will likely stimulate further expansion of the home market.

Research Challenges

Several areas of experiential computing would benefit from additional research. Much of this work would be applicable to both defense and entertainment applications of experiential computing technology. Technologies for image generation, tracking, perambulation, and virtual presence are of interest to both communities, but research priorities tend to be very different. As an example, the factors guiding development of the microprocessors that form the heart of the new Nintendo 64 game machine are very different from those that DOD would have set were it specifying a deployable, low-cost, real-time simulation and training device. For example, the Nintendo system was designed for operation in conjunction with a television and uses an interlaced scanning technique and low-resolution graphics. Most training systems would require higher resolution to enable participants to identify more easily specific features of the environment and to avoid eye strain during periods of extended use and would likely use a progressive scan system similar to most computer monitors. Thus, for military purposes it might be possible to leverage a variant of the Nintendo 64 processor, but the actual processor would probably not do the job.

Image Generation

Visual simulations in defense and entertainment applications share a common need for image generators with a range of capabilities and costs. On the entertainment side, low-cost platforms such as personal computers (PCs) and game boxes, such as those manufactured by Sega or Nintendo, underlie the video games industry. PCs also serve as the primary point of entry to the Internet and therefore are critical to companies providing on-line entertainment, whether through so-called chat rooms or multiplayer games. Larger location-based entertainment centers, such as the flight simulator centers operated by Virtual World Entertainment and the Magic Edge, also are interested in moving away from workstation-based simulators to PC-based simulators as a means of reducing operating costs.

Image generation has long benefited from close linkages between the commercial and defense industries. From its early roots at Evans and Sutherland (E&S) and GE Aerospace, the image generator industry responded largely to defense needs because volumes were low and prices high, typically in the millions of dollars. The high cost limited the use of such simulators outside DOD. Nevertheless, the E&S CT5 (circa 1983) and the GE Compuscene 4 Computer Image generators were benchmarks by which all interactive computer graphics systems were measured for years.

At about the same time, interactive 3D graphics began to migrate into commercial applications. Stanford University Professor James Clark and seven of his graduate students founded Silicon Graphics Inc. to bring real-time graphics to a broad range of applications. Other companies soon followed, creating the now-pervasive commercial market for real-time 3D graphics. As a result, image generation capabilities that cost over $1 million in 1990 are now available on the desktop for one-one thousandth (1/1,000) that price—a drop of over three orders of magnitude in less than a decade. This improvement in price/performance ratios results from both technological advances and a related growth in demand for 3D graphics. By driving up production volumes, increased demand has lowered costs significantly, and the entrance of new competitors into the market has accelerated the pace of innovation and resulted in further declines in cost. As real-time 3D becomes a commodity, the true cost of image generation is switching to software—the time and resources required to model virtual worlds.

As commercial systems become more capable, more opportunities will exist for DOD and the entertainment industry to work together on image generation capabilities, coupling fidelity with the lower costs that stem from producing larger volumes. A number of existing and emerging technologies could potentially be used for DOD training applications. Low-cost 3D image generators exist that can support robust dynamic 3D environments. These range from game machines such as Nintendo 64 to low-cost graphics boards for PCs manufactured by companies such as 3Dfx and Lockheed Martin.

Improvements in low-cost image generators depend on advances in six underlying technologies: processors, 3D graphics boards, communications bandwidth, storage, operating systems, and graphics software. The commercial computer industry will play the leading role in bringing such technologies to the market but will continue to draw from a larger national technology base created by both public and private research programs. Advances in high-end DOD systems may be able to create capabilities that can be used in less expensive systems. Processing power continues to increase with each new generation of microprocessors. Current microprocessors operate at speeds of 200 megahertz or more, and many include multiprocessor logic that can allow several (typically four to eight) processors to work together on a common problem. In the area of 3D graphics boards, some 30 to 40 companies currently offer boards for PCs. As a result, David Clark of Intel Corporation predicts that the performance of graphics chips (the number of polygons generated per second) may double in performance every nine months—twice as fast as processors are improving. Inexpensive chips will soon be able to generate upward of 50 million pixels per second with textures. New communications architectures for PC graphics, such as Intel's accelerated graphics port architecture, will enable over 500 megabytes per second of sustained bandwidth, enabling designers to rapidly transfer texture maps from main memory, thus keeping the cost of 3D graphics low. Because of such advances, producers of PC hardware and software see 3D graphics as a growing application area and are moving quickly to commercialize 3D graphics technology. Both Windows NT and UNIX operating systems support PC-based graphics, and a number of software vendors are porting their applications from the workstation to the PC environment.

Multigen Inc. has announced that it is making products available for Windows NT systems; Gemini Corporation has ported the Gemini Visualization System. Microsoft Corporation's purchase of Softimage, manufacturer of high-end graphics creation software used by both DOD and the entertainment industry, promises to accelerate the graphics capabilities of PCs.

Tracking

One of the areas that has seen insufficient innovation in the past decade, position and orientation tracking, continues to hamper advanced development in experiential computing. Today's tracking systems include optical, magnetic, and acoustic systems. The most popular trackers are AC or DC magnetic systems from, respectively, Polhemus Corporation and Ascension Technologies. These systems have fairly high latency, marginal accuracy, moderate noise levels, and limited range. New untethered tracking systems from Ascension help with the intrusive nature of being wired up but still require the user to wear a large magnet.

Tracking remains a barrier to free-roaming experiences in virtual environments. To meet the goals of the U.S. Army's STRICOM for training dismounted infantry, long tracker range, resistance to environmental effects from light and sound, and minimal intrusion are key to assuring that the tracking does not get in the way of effective training (see position paper by Traci Jones in Appendix D). Similar requirements were expressed at the workshop by Scott Watson of Walt Disney Imagineering. Magnetic tracking is currently used for detecting head position and orientation in Disney's *Aladdin* experience and other attractions, despite the fact that the latency of such systems is roughly 100 milliseconds—long enough to contribute to symptoms of simulator sickness.[3]

As the performance of graphics engines rendering virtual environments increases, the proportional effect of tracker lag is increased. Some optical-based trackers are currently yielding good results but have some problems with excessive weight and directional and environmental sensitivity. Experiments with novel tracking technologies based on tiny lasers are showing promise, but much more work needs to be done before untethered long-range trackers with six degrees of freedom are broadly available in the commercial domain.

While untethering the tracker is a current next-step goal, the ideal tracker would not only be untethered but also unobtrusive. Any device that must be worn or held is intrusive, as it intrudes on the personal space of the individual. All current tracking systems suffer from this problem except for some limited-functionality video tracking systems.

Video recognition systems are typical examples of unobtrusive trackers, allowing users to be tracked without requiring them to wear anything (except for the University of North Carolina video tracker, which actually had users wear cameras!). While this is an ideal, it is difficult to effectively implement and thus has seen only limited application. Some examples include Myron Krueger's VideoPlace and Vincent John Vincent's Mandala system.

Perambulation

Improved technologies are also necessary for supporting perambulation in virtual environments. The U.S. Army's STRICOM has funded the development of an omni-directional treadmill to explore issues associated with implementing perambulation in virtual environments, a topic that is applicable to entertainment applications of VR as well. Allowing participants in a virtual environment to wander around, explore, and become part of a story would greatly enhance the entertainment value of the attraction. It would also enable residents of a particular neighborhood to wander around synthetic re-creations of their neighborhoods to see how a proposed development nearby would affect their area, from a natural perspective and with a natural user interface. Research is needed to improve current designs and to create perambulatory interfaces that allow users to fully explore a virtual environment with floors of different textures, lumps, hills, obstructions, and other elements that cannot easily be simulated using a treadmill.

Technologies for Virtual Presence[4]

Virtual presence is the subjective sense of being physically present in one environment when actually present in another environment.[5] Researchers in VR have hypothesized the importance of inducing a feeling of presence in individuals experiencing virtual environments if they are to perform their intended tasks effectively. Creating this sense of presence is not well understood at this time, but among its potential benefits may be (1) providing the specific cues required for task performance, (2) motivating participants to perform to the best of their abilities, and (3) providing an overall experience similar enough to the real world that it elicits the conditioned or desired response while in the real world. Several technologies may contribute to virtual presence.

- *Visual stimulus.* This is the primary means to foster presence in most of today's simulators. However, because of insufficient consideration of the impact of granularity, texture, and style in graphics rendering, the inherent capability of the available hardware is not utilized to the greatest effect. One potential area of collaboration could be to investigate the concepts of visual stimulus requirements and the various design approaches to improve graphics-rendering devices to satisfy these requirements.

- *Hearing and 3D sound.* DOD has initiated numerous efforts to improve the production of 3D sound techniques, but it has not yet been effectively used in military simulations. Providing more realistic sound in a synthetic environment can improve the fidelity of the sensory cues perceived by participants in a simulation and help them forget they are in a virtual simulated environment.
- *Olfactory stimulus.* Smell can contribute to task performance in certain situations and can contribute to a full sense of presence in a synthetic environment. There are certain distinctive smells that serve as cues for task initiation. A smoldering electrical fire can be used to trigger certain concerns by individuals participating in a training simulator. In addition, smells such as that of hydraulic fluid can enhance a synthetic environment to the extent that it creates a sense of danger.
- *Vibrotactile and electrotactile displays.* Another sense that can be involved to create an enhanced synthetic environment is touch and feel. Current simulator design has concentrated on moving the entire training platform while often ignoring the importance of surface temperature and vibration in creating a realistic environment.
- *Coherent stimuli.* One area that has not received much research is the required coherent application of the above-listed stimulations to create an enhanced synthetic environment. Although each stimulation may be valid in isolation, the real challenge is the correct level and intensity of combined stimulations.

Electronic Storytelling

Part of making a simulated experience engaging and realistic has nothing to do with the fidelity of the simulation or the technological feats involved in producing high-resolution graphics and science-based modeling of objects and their interactions. These qualities are certainly important, but they must be accompanied by skilled storytelling techniques that help participants in a virtual environment sense that they are in a real environment and behave accordingly. "The problem we are trying to solve here is not exactly a problem of simulation," stated Danny Hillis at the workshop. "It is a problem of stimulation." The problem is to use the simulation experience to help participants learn to make the right decisions and take the right actions.

The entertainment industry has considerable experience in creating

SETTING A COMMON RESEARCH AGENDA   41

simulated experiences—such as films and games—that engage participants and enable them to suspend their disbelief about the reality of the scenario. These techniques involve methods of storytelling, of developing an engaging story and using technical and nontechnical mechanisms to enforce the emotional aspects. As Danny Hillis observed:

> If you want to make somebody frightened, it is not sufficient to show them a frightening picture. You have to spend a lot of time setting them up with the right music, with cues, with camera angles, things like that, so that you are emotionally preparing them, cueing them, getting them ready to be frightened so that when you put that frightening picture up, they are startled.

Understanding such techniques will become increasingly important in applications of modeling and simulation in both DOD and the entertainment industry. Alex Seiden of Industrial Light and Magic observed at the workshop that "any art, particularly film, succeeds when the audience forgets itself and is transported into another world." The technology used to create the simulation (such as special effects for films) must serve the story and be driven by it.

DOD recognizes the importance of storytelling in its large-scale simulations. Judith Dahmann of DMSO noted that DOD prepares participants for simulations by laying out the scenario in terms of the starting conditions: Who is the enemy? What is the situation? What resources are available? However, DOD may be able to learn additional lessons from the entertainment industry regarding the types of sensory cues that can help engender the desired emotional response.

Selective Fidelity

One of the primary issues that must be considered in both entertainment and defense applications of modeling and simulation technology is achieving the desired level of fidelity. How closely must simulators mimic the behavior of real systems in order to make them useful training devices? Designing systems that provide high levels of fidelity can be prohibitively costly, and, as discussed above, the additional levels of fidelity may not greatly improve the simulated experience. As a result, simulation designers often employ a technique called *selective fidelity* in which they concentrate resources on improving the fidelity of those parts of a simulation that will have the greatest effect on a participant's experience and accept lower levels of fidelity in other parts of the simulation.

Developers of DOD's Simulator Networking (SIMNET) system, a distributed system for real-time simulation of battle engagements and war games, recognized that they could not fool trainees into actually believing they were in tanks in battle and put their resources where they thought they would do the most good.[6] They adopted an approach of selective fidelity in which only the details that proved to be important in shaping behavior would be replicated. Success was measured as the degree to which trainees' behavior resembled that of actual tank crews. As a result, the inside of the SIMNET simulator has only a minimal number of dials and gauges; emphasis was placed on providing sound and the low-frequency rumble of the tank, delivered directly to the driver's seat to create the sense of driving over uneven terrain. Though users initially reported dismay at the apparent lack of fidelity, they accepted the simulator and found it highly realistic after interacting with it.[7]

The entertainment industry has considerable experience in developing systems that use selective fidelity to create believable experiences that minimize costs. Game developers constantly strive to produce realistic games at prices appropriate for the consumer market. They do so by concentrating resources on those parts of their games most important to the simulation. After realizing that game players spent little time looking at the controls in a flight simulator, for example, Spectrum HoloByte shifted resources to improving the fidelity of the view out the window.[8] Experiments have shown that even in higher-fidelity systems the experience can be improved by telling a preimmersion background story and by giving participants concrete goals to perform in virtual environments.[9]

Selective fidelity is important in both defense and entertainment simulations, though it can be applied somewhat differently in each domain to reflect the importance given to different elements of the simulation. For DOD, selective fidelity is typically used to ensure realistic interactions between and performance of simulated entities, sometimes at the expense of visual fidelity. Hence a DOD simulation might have a radar system with performance that degrades in clouds and rain or an antitank round that inflicts damage consistent with the kind of armor on the target, but it might use relatively primitive images of tanks and airplanes if they are not central to the simulation. The entertainment industry tends to place greater emphasis on visual realism, attempting to make simulated objects look real, while relaxing the fidelity of motions and interactions. An entertainment simulation is more likely to use tanks that look real, but that do not behave exactly like real tanks: their motion may not slow when they travel through mud, or their armor may not be thinner in certain places than in others.

Such differences limit the ability of defense and entertainment systems to be used in both communities. For example, while many modern video games create seemingly realistic simulations, they do not necessarily model the real world accurately enough to meet defense requirements. Granted, there is a genre of video games that strive to be as realistic as possible. Games like *Back to Baghdad* and *EF2000* are popular in large part because they strive for high degrees of accuracy. In addition, because of the long lifetimes of some trainers at DOD, several modern video games far exceed the accuracy of some older operational simulators. But game designers must often break with reality in order to meet budgetary and technological constraints. As Scott Randolph of Spectrum HoloByte noted at the workshop,

> There is always the tendency to do things like take an [infrared] sensor that is good for 10 miles and works best at night. But because you don't want to keep track of [the] time of day, you make it so it always works the same. It can't really see through a cloud or dust, but since there aren't very many clouds in the game and you don't want to keep track of dust, you very quickly end up with a sensor model that doesn't model the real sensor.

While the software in these applications may have the technical underpinnings to produce training devices, their primary goal is entertainment, not accuracy. Given the verification, validation, and accreditation requirements that must be met for DOD training applications and the profit expectations of the entertainment industry, it appears unlikely that a common software application could be written to meet the needs of both communities. This observation does not suggest that DOD and the entertainment industry cannot develop a common architecture or framework (such as network protocols and database formats) for simulation that both communities could use, as is described in the "Standards for Interoperability" section of this chapter.

DOD and the entertainment industry may be able to benefit from their complementary approaches to selective fidelity. The entertainment industry, with support from other industries, will continue to pursue techniques for enhancing visual fidelity. Much of the basic research to support such efforts is being conducted in universities. The National Science Foundation, for example, is funding a Science and Technology Center for Computer Graphics and Scientific Visualization that includes participants from the computer graphics programs at Brown University, the California Institute of Technology, Cornell University, the University of North Carolina at Chapel Hill, and the University of Utah. The center has a long-term research mission (11 years) to help improve the scientific bases for the next generation of computer graphics environments (both hardware and software). Its research focuses on modeling, rendering, interaction, and performance. In contrast, DOD may have a greater incentive to explore ways of incorporating scientific and engineering principles into its simulations to enable entities to behave and interact more realistically. Once developed, techniques for fidelity may be able to be shared between the two communities. One committee member stated that many game developers visit his lab seeking physics-based models for vehicles in a simulated environment.

NETWORKED SIMULATION

Applications

Networked simulation, which allows multiple participants connected to a common network (whether a local-area network or the Internet) to interact simultaneously with one another, is becoming increasingly important to both DOD and the entertainment industry. Both share a common need for adequate network infrastructure to support growing numbers of participants. DOD's goal is to develop a networked training environment in which military operations can be rehearsed with large numbers of participants while avoiding expenditures on fuel, machines, and travel. Participants can range into the thousands or tens of thousands and include soldiers at workstations with weapons-system-specific interfaces, soldiers at keyboards, or computer-generated forces that mimic human interaction. Such large-scale networked simulations are carefully planned and set up, just like actual military maneuvers with real equipment; they are coordinated with radios and include the full range of participants needed to support military operations.

For the games industry the goal of networked games is to provide a shared compelling entertainment experience for participants. Players in such networked systems are most often at the consoles of home computers or at location-based entertainment centers and are connected via local-area networks or the Internet. Internet-based games are an area of strong growth. Currently, Internet gaming supports multiplayer versions of existing computer games that have been modified to allow "Internetworking." Most use proprietary protocols to exchange information and can support interoperability among players using the same game. Though they are still at a simple stage, connecting only tens of players, such games are moving toward larger-scale connectivity. If the number of participants in networked games grows as large as DOD simulations (the targeted size of military simulations has increased by nearly two orders of magnitude over the past decade), new architectures may be required to keep the games from running so slowly that delays (or latency) become perceptible to players.

Research Challenges

Both DOD and the entertainment industry anticipate large growth in the number of participants who engage simultaneously in networked simulations. DOD has already demonstrated systems linking thousands of players and would like to link hundreds of thousands. Most networked games currently allow 8 to 32 players, but as the offerings expand, games could see hundreds or even thousands of networked participants.[10] Such increases in scale pose a number of challenges that will need to be resolved. More participants implies an increase in the size and complexity of virtual worlds. Such requirements, combined with the increased amount of information that must be exchanged among participants, place additional demands on available bandwidth. Greater network traffic implies greater delays in delivering messages along the network unless improved means can be found of designing the network or distributing processing.

Overcoming Bandwidth Limitations

The growth in the number of participants in networked simulations and the desire to share greater amounts of information place increasing demands on bandwidth and computational power of simulation and game systems. Attempts to overcome bandwidth limitations have tended to concentrate on one of two areas: (1) increasing the bandwidth available for networked simulations and (2) minimizing the demand for bandwidth made by networked simulations. Workshop participants agreed that there would be some value in bringing these two communities together to exchange implementation ideas and techniques.

Expanding Available Bandwidth. To overcome bandwidth limitations, both the defense modeling and simulation community and providers of Internet-based games have attempted to develop or acquire greater bandwidth for their systems. DOD has constructed its own network, the Defense Simulation Internet (DSI), to allow simulation systems at distant sites to engage each other. The system in effect provides DOD users with a dedicated wide-area network that is logically separated from the Internet and keeps DOD messages free from other traffic on the Internet. Local-area networks tied to DSI are connected via a T-1 line, which allows high-speed transfers of data. As such, DSI makes each participating site into a big local-area network.

Networked game companies have also attempted to separate their message traffic from that of the Internet to improve reliability and expand available bandwidth. Such companies have typically negotiated with Internet service providers to pay for premium service with certain guarantees of available bandwidth in order to reduce network latency.

Continued improvements in the bandwidth available for networked simulation will continue to derive from advances *outside* the defense modeling and simulation community or the entertainment industry. Telecommunications companies and Internet service providers continue to upgrade the capacity of their networks and network connections to provide higher-speed access and greater bandwidth. DOD will continue to expand the capacity of its networks for command, control, communications, computing, and intelligence (C^4I) and the Defense Simulation Internet. To the extent that DOD's C^4I community becomes more closely linked with the commercial communications industry, some of the existing constraints on bandwidth may be relaxed.[11]

Reducing Bandwidth Requirements. Other attempts at overcoming bandwidth limitations have focused on using existing bandwidth more efficiently and reducing the amount of bandwidth demanded by networked simulations. Networked game companies, cognizant that most players access the Internet via 14.4- or 28.8-kilobits-per-second modem connections, are striving to customize their network data to reduce data transmission requirements while maintaining the entertainment value of their applications. Military simulation designers have paid relatively little attention to determining which data transmissions can be dispensed with while retaining acceptable reality at the application level. The military could leverage game developers' expertise in determining what data reduction might be achievable through techniques developed by the entertainment industry. DOD and the entertainment industry could also become more involved in work that is under way internationally in the computer and consumer electronics industries to develop image compression technologies and standards. To date, neither DOD nor the entertainment community has been heavily involved in MPEG-4.

DOD and the Internet community have pursued another possible solution to the bandwidth problem: multicast routing systems that incorporate software-based area-of-interest managers (AOIMs) to direct packets of information across a network to particular groups of listeners.[12] Such systems allow any member of a group to transmit messages (containing text, voice, video, and imagery) to all other members of the group via a single transmission.[13] This approach prevents the sender from having to transmit individual copies of the message to all intended recipients, freeing resources for other purposes. Machines that are not part of the group ignore the packet at the network interface, eliminating any need for the central processing unit to read the packet. Proposed partitioning schemes are based on spatial (geographic groupings based on locality), temporal (e.g., real-time versus nonreal-time), and functional (e.g., voice communications, aircraft) characteristics. AOIMs distribute partitioning algorithms among hosts rather than rely on a central AOIM server. Used in conjunction with multicast routing, they can help minimize the amount of bandwidth needed to support networked simulations.

Work on multicast routing and AOIMs has been ongoing for several years. The Naval Postgraduate School has incorporated multicast into its NPSNET for experiments across the Internet. DMSO has also invested in the development of AOIM filters as part of its High-level Architecture (HLA), and the HLA's Run-Time Infrastructure (RTI) will soon support a comprehensive AOIM capability.[14] The Internet Engineering Task Force (IETF) has also initiated an effort to implement multicast mechanisms to support distributed simulation. Its Large Scale Multicast Applications working group has developed documentation to describe how IETF multicast protocols, conference management protocols, transport protocols, and multicast routing protocols can support large-scale distributed simulations, such as DOD simulations containing 10,000 simultaneous groups and upward of 100,000 virtual entities.[15]

Nevertheless, additional research is needed to expand the capabilities of AOIMs beyond those of simple filters and to make them generalizable across problem domains. The IETF, for example, has identified seven areas in which existing Internet protocols are insufficient to support large-scale distributed simulation networks (see Table 2.1). Among these is the need to develop multicast protocols that can provide the quality of service needed for distributed simulation: different types of messages must be transmitted with different degrees of reliability and latency. In addition, research is needed to (1) help define a network software architecture that properly uses AOIMs; (2) determine how best to program AOIMs and how to generalize the concept of AOIMs so that as applications change new AOIMs can be downloaded to match the application; and (3) implement and test AOIMs on a wide-area network basis.

Network Latency[16]

Latency is a major barrier to fast-action Internet games and to large-scale simulations generally. To make participants feel as though the system is responding in real time, designers of fast-action games typically try to keep the delay time between the moment players instruct their simulators to take certain actions (e.g., fire a gun, change direction) and the time the system generates the appropriate response to 33 milliseconds or less. Maintaining such latencies across large distributed networks, such as the Internet, is difficult. Electrical and optical signals transmitted across such networks can experience several types of delays. The fundamental limitation is the speed of light: signals cannot travel roundtrip in fiber optic cable between New York and San Francisco, for example, in less than about 54 milliseconds. But signals encounter additional delays as they travel through large networks: modems must format messages for transmission over the network; routers must determine

TABLE 2.1 Additional Capabilities Identified by the Internet Engineering Task Force That Are Needed to Support Multicast in Distributed Interactive Simulations

Need	Description
Resource reservation in production systems	The capability to reserve a specified amount of network bandwidth for a given simulation to help ensure that certain messages can be transmitted with higher levels of service. The proposed Resource Reservation Protocol is one candidate for this need.
Resource-sensitive multicast routing	Routing protocol that determines the paths of packets through the network based on the relative congestion of different pathways and the quality-of-service demands of different message types.
Multicast capabilities that take advantage of all multicast-capable data link protocols	Multicast capabilities need to be extended to different types of wide-area networks, such as those that use asynchronous transfer mode communications.
Hybrid transmission protocols	A set of transmission protocols that can provide the range of quality-of-service and latency requirements of distributed interactive simulations, such as best-effort multicast of most data, reliable multicast of critical reference data, and low-latency reliable unicast of data among arbitrary members of a multicast group.
Network management for distributed systems	A protocol for managing network resources, such as the Simple Network Management Protocol used on the Internet.
Session protocols to start, pause, and stop distributed simulations	Procedures and protocols that facilitate coordinated starting, stopping, and pausing of large distributed exercises.
Integrated security architecture	Mechanisms to ensure the integrity, authenticity, and confidentiality of communications across the network.

SOURCE: Pullen, J.M., M. Myjak, and C. Bouwens. 1997. "Limitations of Internet Protocol Suite for Distributed Simulation in the Large Multicast Environment," a draft report of the Internet Engineering Task Force dated March 24; available on-line at ftp.ietf.org/internet-drafts/draft-ietf-lsma-limitations-01.txt.

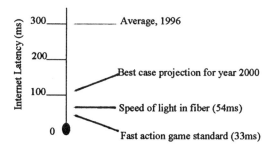

FIGURE 2.1 Estimated latency for round-trip coast-to-coast Internet transmissions. SOURCE: Position paper prepared for this project by Will Harvey; see Appendix D.

how to send the messages through the network. Queuing delays that are due to congestion on large networks and packet losses (which require messages to be re-sent) can add significantly to the latency of distributed networks, especially public networks like the Internet that handle large volumes of traffic.

As a result, even premium services over the Internet cannot generally meet the latency requirements for fast-action simulations conducted across a widely distributed network. According to Will Harvey of Sandcastle Inc., network latency for roundtrip coast-to-coast transmissions across the Internet will not drop below 100 to 130 milliseconds by the year 2000 (Figure 2.1).[17] Fast-action simulations do not operate realistically with such latencies. Participants trying to dodge bullets may become frustrated because the response time is too slow for them to dodge, or they may feel cheated because the program displays their character such that it appears they have dodged it when they have not.

Latencies are highly dependent on the architecture of distributed networks. In a lockstep architecture (Figure 2.2) each individual machine controls all objects locally and broadcasts changes to the other machines via a central server. If a player pushes a button to make his or her character jump, for example, the player's own machine will update the position of the character and send a message to the server indicating that the jump button was pressed. The server will broadcast a message to other players' machines that the first player pressed the jump button, and those player's machines will update the first player's position accordingly. The simulation advances one cycle when each machine has received a complete set of user input from all participating machines. Since advancing a cycle requires complete exchange of user input, the responsiveness of the system is limited by the latency of the slowest communication link and is contingent on the reliability of all nodes.

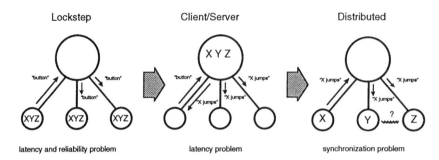

FIGURE 2.2 Alternative architectures for distributed simulations. NOTE: In these pictures the large circle is a server or multicast router in a building. The small circles are machines in people's homes. X, Y, and Z represent objects controlled by users from their own homes. Proxies not shown in these figures display the objects on every machine. The X, Y, and Z letters represent the point of control of each object. SOURCE: Position paper prepared for this project by Will Harvey; see Appendix D.

In a client/server architecture each machine independently sends its user input or action request to a central server, which then relays the information to each player's client machine. For example, if a player pushes the jump button, the client machine will send a message to the server indicating that the button has been pushed. The server will then update the position of the player's icon and send the updated position to all client machines participating in the game. Controlling an object from a client machine still entails a roundtrip delay (from client to server and back), but the responsiveness of any individual client machine is not affected by the communication speed—or reliability problems—of the other machines.

In a fully distributed architecture, machines control objects locally and broadcast the results of actions to other machines, which receive the information with some time delay. If a player presses the jump button, his or her machine will update the position of his or her character and send the updated information, via the central server, to each of the other players engaged in the game. Each machine has immediate responsiveness controlling its own objects but must synchronize interactions between its own objects and other objects controlled by remote machines.

In the lockstep and client/server architectures, responsiveness is limited by the roundtrip communication latency to the server. In the distributed architecture, responsiveness is not limited by latencies because objects are controlled locally and the players' own objects respond to commands with little delay; however, this architecture creates problems of synchronization. Players' own objects are displayed in near real time, but other players' objects are displayed with a delay equal to the latency of the system. While such synchronization problems are not of concern if interactions between objects are minimal, they can create problems with shared objects.

Attempts to resolve latency and synchronization problems can take several approaches. The first is to improve the speed of the underlying network. Work is under way to develop and deploy a new algorithm for queue management called Random Early Detection that will help minimize queuing delays across the Internet and other networks.[18] Other research is investigating ways to speed the delivery of time-sensitive packets across the Internet by establishing different levels of service quality.[19] For a price premium, users will be able to designate that their packets need to be delivered with minimal delay. Similar performance is available today through the use of dedicated networks that some Internet gaming companies and the DOD have built to support distributed simulations. Such networks avoid delays caused by congestion in public networks. Nevertheless, all such efforts in this area cannot reduce latencies below those imposed by the speed of light itself.

Other attempts at improving responsiveness recognize the latencies inherent in distributed systems and attempt to compensate for them by predicting the future location of objects, a technique called *dead reckoning*. This technique accommodates the delay with which information is received from other participants by predicting their actions using information such as position, velocity, and acceleration and by bringing all objects displayed on each simulator into the same time frame. Such techniques are effective only insofar as the motions or actions of objects are predictable and continuous; they cannot yet anticipate future changes in the course of objects (although future research may allow the development of more sophisticated algorithms that can anticipate deviations from continuous motions, perhaps by incorporating information about terrain features or past flight trajectories). Participants in simulated tank engagements have found ways to outsmart such dead-reckoning techniques: before passing in front of an enemy tank, they will accelerate quickly and then stop abruptly so that the enemy tank will incorrectly predict and display their position.

Alternatively, some researchers are developing techniques for synchronizing events across distributed simulations.[20] Such approaches assume that information from remote simulators will always be received with a time delay and that many actions cannot be predicted accurately. Thus, they show objects from remote machines with an inherent time delay. If users have no interactions with remote objects, the time delay does not interfere with the simulation and the system need not compensate for the time differential. But if objects from local and remote machines do interact, synchronization technologies compensate for the time difference. These technologies give users the impression that the network has zero latency, or immediate responsiveness, but they do so by sacrificing some degree of fidelity. For example, a 100-meter car race may actually be a 90-meter race with 10 meters of compensation interjected at the appropriate places to synchronize the outcome; similarly, a ball thrown from one player to another may travel faster when moving away from than toward the local player (Box 2.1). While having demonstrated some efficacy in game applications, additional analysis will be needed to determine the suitability of such techniques for defense simulations that require high levels of fidelity.

STANDARDS FOR INTEROPERABILITY

A related area of interest to DOD and the Internet games community is standards for interoperability. Interoperability is the ability of various simulation systems to work with each other in a meaningful and coherent fashion.[21] It is often defined as a matter of degree: a simulation is considered *compliant* if it can send and receive messages to and from other simulators in accordance with an agreed-upon specification. Two or more simulators are considered *compatible* if they are compliant and their models and data transmissions support the realization of a common operational environment. They are *interoperable* if their performance characteristics support the fidelity required for the exercise and allow a fair contest between participants in which differences between individual simulators are overwhelmed by user actions. For example, a flight simulator with no damage assessment capability would not be interoperable with other simulators having this capability—even if they can communicate data effectively—because it could not detect that it had been hit by an enemy missile and destroyed.

Achieving interoperability between simulation systems requires (1) a common network software architecture with standard protocols that govern the exchange of information about the state of each of the participants in the simulation; (2) a common underlying architecture for maintaining information about the state of the environment related to a particular simulator; and (3) a common representation of the virtual environment.[22] As the size and scale of defense simulations grow, participants will need a consistent view of the battlefield so that they can agree on the location of objects there and on the timing of events. Given that different players will have different ways of gathering information (e.g., radars and other sensors as well as information relayed from command

> **BOX 2.1**
> **An Example of Event Synchronization**
>
> Consider a football game played by multiple players in a distributed network. It is important that the football be controlled locally on the machine of the player currently holding the ball so that latencies are minimized; otherwise, the ball will not move at the same time as the player's hand, and a lag may be apparent. If the football is passed to another player, it must then be controlled by the other player's machine, but passing the football is complicated by the fact that the positions of the players on the two computers may not be synchronized because of latencies in the network.
>
> The solution to this problem rests on two observations. First, by *migrating* the football from one machine to another, the latency problem can be corrected. Second, two players do not need to see exactly the same thing on their two screens, as long as they agree on the outcome and no one feels cheated. So if one player throws another a football, the football can travel slightly slower from the first to the second player on one screen than on the other, giving control of the football a chance to migrate from one computer to the other midway through the path. Neither player can tell the difference.
>
> To demonstrate the viability of this approach, Sandcastle Inc. developed an experimental Internet ping pong game in which two players hit a ball back and forth to each other, controlling individual paddles. Each player sees his or her own paddle in real time and the opponent's paddle with some time delay. Each player sees the ball travel away from his or her paddle at a slower speed than the ball travels toward his or her paddle, so that at the point at which the ball reaches the opponent's paddle, the player sees the ball and the opponent's paddle at a point in the past equal to the latency of the system. In experiments, participants could not detect a delay in the system, even with network latencies of two-thirds of a second (670 milliseconds).
>
> ---
> SOURCE: Will Harvey, Sandcastle Inc.

and communications centers), they do not need to have the *same* information, but the simulation itself should not alter the information they receive. Meeting these requirements allows tank simulators, for example, to be designed and interconnected so that their operators can share information and train jointly in a common virtual battlefield.

Related to this capability is *composability*, the ability to build simulations using components designed for other simulations. Composability is a significant concern for DOD, which cannot construct a single integrated simulation that serves all possible purposes and would like to minimize the development costs of new simulations. Using composable simulations, for example, a simulation designed to train aircrews and ground forces in conducting close air support operations could be built using simulated aircraft and simulated soldiers that were designed for other simulations. Ensuring this type of interoperability requires a common architecture for the design of simulations and a common understanding of the types of tasks conducted by the individual simulators and those conducted by the integrating system. While most existing simulations were designed for a particular purpose and may not be able to be combined into larger simulations, future systems may be able to be designed in ways that will allow greater interoperability and composability.

DOD Efforts in Interoperability

Over the past several years DOD has attempted to develop standards to promote interoperability among its simulation systems. Such efforts have needed to accommodate the heterogeneity and scale of entities modeled in the virtual battlefield—combat aircraft, tanks, and ships, and refueling vehicles—and support the full generality of participant interactions. To this end the defense simulation community developed standards, such as Distributed Interactive Simulation (DIS) standards, that aim primarily at achieving "plug-and-play" interoperability among simulators developed by independent manufacturers.

DIS is a group of standards developed by members of the defense modeling and simulation community (both industry and university researchers) to facilitate distributed interactive simulations. They can be used for hosting peer-to-peer multiuser simulations in which objects (typically vehicles) move independently, shoot weapons at each other, and perform standard logistics operations such as resupply and refueling. The DIS protocols include a variety of industry and military standards for computer equipment and software, as well as the Transmission Control Protocol/Internet Protocol (TCP/IP) networking protocols used over the Internet. Specific protocols have had to be devised to define the communications architecture for distributed simulations as well as the format and content of information exchanges, the types of information relevant to entities (such as tanks, aircraft, and command posts) and the types of interactions possible between them, simulation management, performance measures, radio communications, emissions, field instrumentation, security, database formats, fidelity, exercise control, and feedback.[23] DIS has a well-developed simulation management subprotocol for setting up and controlling individual players in an exercise. Consequently, DIS can achieve certain levels of data interchange for one-on-one and unit-level interactions but cannot support aggregation and disaggregation of units.

The DIS standard is derived from protocols developed for DOD's Simulation Network (SIMNET) system, adopting its general principles, terminology, and protocol data unit (PDU) formats for transmitting information between simulators.[24] The initial set of protocols (DIS 1.0) was accepted by the standards board of the Institute of Electrical and Electronics Engineers (IEEE) in 1993 and is now codified under IEEE standard 1278-1993. These protocols were subsequently recognized by the American National Standards Institute.

In addition, DOD has been pursuing development of the High-level Architecture (HLA) to facilitate both interoperability and composability. HLA is a software architecture that defines the division of labor between simulators and a layer of support software, called the Run-time Infrastructure (RTI), that facilitates interoperability. It consists of specifications, interfaces, and standards for a broad range of simulations, from combat simulations to engineering analyses. Groups of people wanting to establish interoperability among their simulators via the RTI do so by creating a *federation*. Federation members make their own decisions about the types of entities that will be included in simulations and the types of information they will exchange. Individual simulators post state-change information to the RTI and receive state-change information from other simulators via the RTI. The RTI makes sure all parties to a federation receive state-change data and ensures that the federation's data are time synchronized and routed efficiently to the other simulations in the federation.

Development of the HLA is being managed by the Architecture Management Group, which is chaired by DMSO and has representatives from all military services and DOD agencies developing advanced modeling and simulation systems. Members represent a wide range of military applications, from training and military operations to analysis, test and evaluation, and engineering-level models for system acquisition and production. Development of the HLA was initiated in 1994, when DARPA awarded three contracts for the definition of a high-level architecture for advanced distributed simulations. Contractors analyzed the needs of prototype federations in four areas: (1) platform simulators (such as an M-1 tank simulator and Close Combat Tactical Trainer; (2) joint training (simulation-like war games that occur in accelerated time); (3) analytical simulations of systems to support joint theater-level war-fighting and support activities; and (4) engineering federations to design, test, and evaluate new military systems. The final briefings from these contractor teams were received in January 1995, and a core team of individuals synthesized the inputs, with additional insight from other ongoing DOD modeling and simulation programs to arrive at the initial definition of the HLA.[25] Testing of prototype federations was completed in July 1996 to determine if the RTI was broadly enough defined to be useful across a wide range of federations. Test results informed development of the HLA Baseline Definition, which was completed in August 1996 and approved by the Under Secretary of Defense for Acquisition and Technology as the standard technical architecture for all DOD simulations in September 1996. All simulations developed after October 1, 1999, must comply with HLA; no existing simulations that are not compliant with HLA may be used after October 1, 2001, unless they are converted to the standard.[26]

BOX 2.2
The Processes Used to Create Standards

The development of a standard for distributing 3D computer graphics and simulations over the Internet has taken the quick path from idea to reality. In 1994 Mark Pesce, Tony Parisi, and Gavin Bell combined their efforts to start the VRML effort. Their intention was to create a standard that would enable artists and designers to deliver a new kind of content to the browsable Internet.

In mid-1995 VRML version 1.0 emerged as the first attempt at this standard. After an open Internet vote, VRML 1.0 was to be based on Silicon Graphics Inc.'s (SGI) popular Open Inventor technology. VRML was widely evaluated as unique and progressive but still not useable. At this point, broad industry support for VRML was coalescing in an effort to kick-start a new industry. Complementary efforts were also under way to deliver both audio and video over the Internet. The general feeling was that soon the broad acceptance of distributed multimedia on the Internet was a real possibility and that VRML would emerge as the 3D standard.

After completion of the VRML 1.0 standard, the VRML Architecture Group (VAG) was established at the Association for Computing Machinery's Special Interest Group on Computer Graphics (SIGGRAPH) annual conference in 1995. It consisted of eight Internet and 3D simulation experts. In early 1996 VAG issued a request for proposals on the second round of VRML development. The call was answered by six industry leaders. The selection of the VRML 2.0 standard was made via open voting and occurred in a short time frame of about two weeks. SGI emerged as the winner with its "Moving Worlds" proposal. By this time over 100 companies had publicly endorsed VRML, and many of them were working on core technologies, browsers, authoring tools, and content. At SIGGRAPH '96, VAG issued the final VRML 2.0 specification and made a number of other significant announcements.

DOD hopes that the HLA will be adopted outside the defense modeling and simulation community. To facilitate this process, RTI software will be made freely available as a starter kit.[27] The software will be in the public domain; initial release will be for Sun workstations, with later releases for Silicon Graphics, Hewlett-Packard, and the Windows NT platforms. Adoption of HLA beyond DOD is questionable, however; members of the entertainment industry noted at the workshop that HLA's development took place largely without their input—or that of other non-defense communities. As a result, many representatives of the games community believe that HLA will not meet their needs; at the workshop

---

To help maintain VRML as a standard, VAG made several concrete moves. First, it started the process of creating the VRML Consortium, a not-for-profit organization devoted to VRML standard development, conformance, and education. Second, VAG announced that the International Organization for Standardization would adopt VRML and the consensus-based standardization process as its starting place for an international 3D metafile format.

Like VRML, DIS standards were generated in an open process via the biannual Workshop on Standards for the Interoperability of Distributed Simulations. Though the first version of DIS derived largely from standards developed by BBN Corporation for the SIMNET program, participation in a revision of the standards was greatly expanded. The first workshop was held in Orlando, Florida, in September 1989 and defined the shape of the DIS standards as they progressed from protocols for DOD's Simulation Networking (SIMNET) program to the version 2.1.4 standards that exist today. Attendance at the workshop grew from 150 people in 1989 to more than 1,500 in September 1996.

In contrast, the HLA design and prototype implementations were developed by a small group of DOD officials and contractors in a more closed fashion that did not solicit input from the modeling and simulation community at large—despite an interest in promulgating the standard broadly. Doing so speeded development of the standard, but significant concerns were expressed that requirements of the broader community were being left out in the rush to completion. Few members of the Communications Architecture Group of the DIS workshop participated in HLA's development, nor did representatives from industries outside defense.

SOURCES: Position paper prepared for this project by Brian Blau; see Appendix D. Also, Lantham, Roy. 1996. "DIS Workshop in Transition to . . . What?", *Real Time Graphics* 5(4):4-5.

many indicated that they were not even aware that the standard was being developed.[28] Further, the process used to develop HLA may have offended those who were not involved in the development process.[29] HLA was not developed in as open a manner as DIS standards and standards for the Internet community, such as the Virtual Reality Modeling Language (VRML) (Box 2.2).

Interoperability in the Entertainment Industry

The entertainment industry, to date, has expressed different interests regarding interoperability standards. While DOD has a strong interest in ensuring that various simulation systems can work together, the enter-

---

BOX 2.3
Virtual Reality Modeling Language

VRML is an evolving standard used to extend the World Wide Web to three dimensions. It is a widely accepted commercial standard, one that people heavily involved with the Internet are seriously thinking about and adopting for their software development. The current version, VRML 2.0, is based on an extended subset of the Silicon Graphics Inc.'s (SGI) Open Inventor scene description language. It has both a static component and an interactive component. The static component features geometry, textures, 3D sound, and animation. The interactive component contains a flexible programmability that can be added to a VRML file through the use of Java code, another Internet technology. This addition of Java allows not only graphics data to be exchanged with VRML but also behaviors. VRML is both comprehensive and unfinished, with its current draft exceeding several hundred pages.

Complex issues surrounding real-time animation in VRML 2.0 include entity behaviors, user-entity interaction, and entity coordination. Many factors are involved. To scale to many simultaneous users, peer-to-peer interactions are necessary in addition to client-server query and response. An approved specification for internal and external behaviors is nearly complete. VRML 2.0 will provide local and remote hooks (i.e., an applications programming interface, or API) to graphical scene descriptions. Dynamic scene changes will be stimulated by any combination of scripted actions, message passing, user commands or behavior protocols (such as DIS or Java). Thus, the forthcoming VRML behaviors standardization will simultaneously provide simplicity, security, scalability, generality, and open extensions.

VRML is not yet complete, and there is still ongoing work. Three specific areas being addressed include definition of an external API, definition of

tainment industry places strong emphasis on developing proprietary systems and standards that preclude interoperability. Mattel, for instance, encrypted data output from its Power Glove input device so that it could not be used with competitors' game devices. Nintendo and Sega game machines cannot interoperate with each other or with computer-based video games. A flight simulator game produced by Spectrum HoloByte Inc. cannot be used in a mock fight against Strategic Simulation Inc.'s *Back to Baghdad* game. Even within the multiplayer networked games, each player uses the same game program. Commercial standards have therefore not sought interoperability between independent systems, but have attempted to allow independently produced software titles to integrate with the same user front-end software (such as operating systems,

---

a default browser scripting language, and compression technology. The purpose of the external API is to provide a way for Web developers to write programs that are able to drive a VRML simulator or a VRML browser. A call has been issued requesting proposals for this external API, and the VRML community expects to have that finalized in 1997. The purpose of the default scripting language is to have a standard language that can be embedded in a VRML file, a language understood by every VRML browser. A call has been requested for proposals for this language.

The purpose of the compression technology call is to be able to compress VRML data, hence minimizing bandwidth requirements on the Internet. A particular response to this call is notable in that it is a joint proposal between IBM and Apple Computer. The proposal is for the binary compression of VRML files and is significant because IBM and Apple have decided to open up their patents on geometry compression, providing them free to the Internet. IBM and Apple are providing royalty-free licenses to VRML developers for this compression. This is a significant step in the types of collaboration the Internet environment seems to be bringing out in the commercial world.

The VRML community is taking steps to ensure widespread adoption. The major step in this direction is that a VRML consortium is being formed as a permanent fixture of the Internet community. Additionally, the International Organization for Standardization (ISO) has picked up VRML as the 3D metafile standard, a selection it has been working on for quite a few years. By selecting VRML, ISO changed the way it normally does business. For the 3D metafile standard, ISO put aside its normal process of building its own standard. Instead it opted to adopt VRML as it is today in order to finish quickly. The entire ISO standards process for VRML is expected to be completed within just 14 months, cutting literally 75 percent of the time that it normally takes to create a standard.

Web browsers, or graphics libraries) so that players with different computer systems can play each other. Standards such as VRML 2.0, OpenGL, and DirectX are aimed in this direction (Box 2.3). As a result of these standards, a user can use the same software to run a variety of game applications.

Research Areas

Growing interest in networked simulation suggests that research into standards and architectures for interoperability will continue to be needed. Areas of particular interest include protocols for networking virtual environments, architectures to support interoperability, and interoperability standards.

Virtual Reality Transfer Protocol

Researchers in advanced browser and networked game technologies are beginning to have similar concerns regarding interoperability. The desire from the browser side is for interactive 3D capabilities, but the VRML standard does not have the ability to support peer-to-peer communications—the type of communications required for networked interactive gaming. Browsers allow heterogeneous software architectures to interoperate, thanks to the *http* standard.

Additional work is ongoing toward the development of a standard for networked virtual environments. A 1995 National Research Council report identified networking as one of the critical bottlenecks preventing the creation of large-scale virtual environments (LSVEs) as ubiquitously as home pages on the Web.[30] Numerous important component technologies for LSVEs have been developed, but that work is not yet complete, and significant research and integration work remain.[31] Some of the integration work is to merge the ideas from the LSVE research community with work from the Internet/World Wide Web community. The integration effort discussions are occurring under various titles—the virtual reality transfer protocol and the IETF's large-scale multicast applications working group.[32]

Among the component technologies that have enabled rapid exponential growth of the Web are HTML and *http*. HTML is a standardized page markup language that allows the placement of text, video, audio, and graphics in a nonplatform-dependent fashion. HTML is being extended into four dimensions with VRML via the addition of 3D geometry and temporal behaviors. *http* is the hypertext transfer protocol, an applications layer protocol used to serve HTML pages and other information across the Internet. *http* binds together several dissimilar protocols including the ftp, telnet, gopher, and mailto protocols; hence, it is an integration or metaprotocol. To support LSVEs across the Internet, it is expected that a continuum of dissimilar protocols will have to be integrated into a single metaprotocol, an applications-layer protocol called the virtual reality transfer protocol (*vrtp*).[33] At a minimum, *vrtp* will combine *http* (for URL service), peer-to-peer communications (such as DIS and its relatives), multicast streaming (for audio/video streams), Java agents, heavyweight object server protocols (such as CORBA and ActiveX), and network monitoring.

With *vrtp* and by using the existing Internet infrastructure, Web-scalable LSVEs will be constructible. VRTP will be an open architecture that will use standards-based public-domain software and reusable implementations, all taking advantage of the sustained exponential growth of "Internetworked" global information. The *vrtp* project is ongoing and taking a similar standards approach to that used for VRML, an open, public, Web-based forum for technical discussion and adoption.

Architectures for Interoperability

Outside the VRML/*vrtp* communities, there is little academic research on solving the network software architecture interoperability problem. Nevertheless, the benefits of additional research could be large, and there are many unsolved research problems. The key to solving the network software architecture problem is to understand that doing so in a scalable way will require attention not just to networking issues but also to networking and software architecture issues. This means that one cannot just design an applications-layer protocol to establish message formats. Attempts to design distributed simulations must consider the scarcity of available network bandwidth and processor cycles and must attempt to minimize latencies across the network. Brute-force methods involving large computers can do simple aggregation to help minimize bandwidth, but such methods are expensive and limited. Distributed solutions that work with both network bandwidth and processor cycle limitations are inherently more scalable.[34] Real-time simulation requires high-performance systems and low communications latency. Adding more layers of abstraction and protocol, as is common for achieving interoperability, can work against the need to meet the latency and performance requirements of networked simulations. Research into application-level protocols and architectures must take these considerations into account.

From the networked games perspective, heterogeneity in software architecture is not yet possible, although some research is being done. The desire to build large-scale gaming, on the order of thousands of players in the same virtual world, may require game companies to move toward more malleable standardized protocols. Game companies will have to decide how to define their applications-layer protocols, again the same thing that everyone in the large-scale virtual world must be able to do. One proposal being considered is for the creation of an applications-layer protocol called GameScript. The purpose behind GameScript is to define a standard applications-layer protocol that allows games designed by different manufacturers to communicate. One of the most often described uses for GameScript is that players in one game see players in neighboring games at the boundaries of their virtual worlds. This serves as a teaser for game players to deposit money and try the next game. Several telephone companies and small venture-capital-funded start-ups are working on this, but no new products have been announced yet. GameScript work is similar to HLA and to the virtual reality transfer protocol work.

Interoperability Standards

Common interoperability standards could have benefits to both DOD and the entertainment industry, enabling them to solve common problems with common solutions. At present, there is no consensus in the games industry on the desirability of a common set of interoperability standards. While some game developers see common standards as a means of facilitating attempts to move networked games onto the Internet, many do not yet consider common standards a high priority. According to Warren Katz of MäK Technologies, resistance to common interoperability standards is generally based on four factors:

- *Technical considerations*—Common standards tend to be designed to accommodate a wide range of potential uses and therefore are not optimal for any particular use. Given existing limitations in bandwidth for Internet-based games (most potential users connect to the Internet via modems that communicate at 14.4 or 28.8 kilobits per second), many game companies prefer to design custom protocols that maximize performance.
- *Not-invented-here syndrome*—Many commercial firms have a bias against technology developed outside their own organization. Engineers in many companies believe they can develop better protocols that will provide a more elegant solution to a problem or that will speed processing times considerable. Acceptance of an existing solution implies that they are incapable of doing better.
- *Strategic value of proprietary solution*—Proprietary networking protocols are viewed as a strategic competitive advantage. Use of a public standard would eliminate one element of advantage by allowing compet-

itors to use the same technology. In addition, use of a public standard could signal that a company is unable to develop a better solution.

- *Control*—Adoption of an industry or public standard reduces the control a company has over its protocols. Standards committees determine changes to the protocol. Companies that control their own protocols can upgrade them at their own pace, as the need arises.

These factors have, to date, stymied the use of DIS standards in the entertainment industry. According to Warren Katz, many video game companies have examined DIS protocols for suitability in their games, but few have implemented them. Some companies have found DIS protocols to be too big and complex, performing operations that were not relevant to video games and slowing the performance of the system.[35] Others view DIS as a standard for military applications and do not consider it appropriate for nonmilitary games. As a result, the DIS protocols have not been embraced in an unmodified form by any game company. Several game companies have developed protocols derived from DIS that include only those functions needed to support their applications. Each of these implementations is proprietary to the developing company and not interoperable with other companies' protocols.

Many companies developing networked games attempt to use their proprietary protocols to their competitive advantage. Proprietary solutions serve as a means of differentiating one company's products from another's and, possibly, of generating revenues through licensing. Should one company develop a set of standards that is perceived as superior in some critical respects (whether allowing greater numbers of participants or by offering better games), it can make its games more attractive to game players. Furthermore, other companies might want to develop games that interoperate with that standard. The standard developer can charge a licensing fee that generates revenues in addition to whatever revenues it makes from selling its games. Use of a common or public standard does not provide such opportunities.

Not all companies will necessarily follow this business model. Some game companies might, for example, see their competitive advantage in the content of their games, not necessarily their networking capabilities. They might attempt to develop a protocol that they license freely in the hope that it will become widely adopted and facilitate growth in the market for networked games, leading to more sales of their titles. Adoption of this model will require a shift in the current business model of most video game companies. Some impetus for this shift may follow from growth in the market for multiplayer Internet games, but none has yet been seen in the games market.

As the Internet games industry matures, it is possible that common standards for interoperability, such as HLA, could become of interest to game developers as a means of allowing them to reuse portions of one simulation to populate another. Anita Jones, DOD's director of defense research and engineering through May 1997, suggested at the workshop that as games become more complex, incorporating more players and a greater number of possible types of interactions between and among players, game developers may shift to a new product development strategy in which they construct games from components already developed for other simulations. Gilman Louie, of Spectrum-HoloByte, suggested that Internet-based games may encourage such change. Much like television or cable, Internet channels will need a great deal of programming with frequent updates and new activities so that users can experience something new to do every time they log on. The code base will need to be designed to allow easy upgrades so that designers can replace particular objects without completely redesigning the system.[36]

Others at the workshop suggested that HLA might be advantageous in allowing real-time simulations to interoperate with simulations that progress faster or slower than real time. In the video games world this capability would mean that a high-level strategy game that typically runs faster than real time could interoperate with a real-time simulator: a turn-based chess game could interoperate with a simulation of the motion of the pieces. Though these capabilities could be very useful in a game environment, it would be unreasonable to assume that any games company would adopt HLA without a strong outside influence. The more likely scenario, according to Warren Katz, is that a game company will adapt desired features of HLA within its own proprietary protocol.

COMPUTER-GENERATED CHARACTERS

One of the major challenges in creating a useful simulation system is populating simulated environments with intelligent characters and groups of characters. While some or even many of the entities present in simulated worlds may be controlled by human operators who are networked into the simulation, many are likely to be operated by the computer itself. Such *computer-generated characters* are a critical element of both defense and entertainment systems and serve a wide range of functions. Computer-generated characters may serve as an adversary against which a game player or user of a training system competes, such as an opponent in a computerized chess game or an enemy aircraft in a flight simulator. At other times they may serve as collaborators that guide participants through a virtual world or serve as a crew member. In large networked simulations, computer-generated characters may control the actions of elements for which human controllers are unavailable, whether a rival tank, a wingman, or a copilot.

Computer-generated Characters in Entertainment

Virtually all sectors of the entertainment industry are interested in computer-generated character technologies as a means of creating more believable experiences for participants and allowing greater automation of services where possible. Companies in the video games, virtual reality, and filmmaking sectors are developing or have deployed products that incorporate computer-generated characters. Since the creation of compelling virtual characters holds such allure, whoever can develop compelling complex characters in a rich fantasy world available through an accessible medium at an affordable price stands to profit handsomely. It is therefore likely that any company that devises a successful approach to solving this problem will be disinclined to share the enabling technologies in order to protect its competitive advantage.

Video Games

Almost every genre of computer games, whether it be sports, action, strategy, or simulation, depends on computer-generated opponents. The ability of a game to attract and entertain players is directly linked to the quality of the computer-generated competitors in a game. All of the best-selling PC games (*Chessmaster, Madden Football, Command and Conquer, Grand Prix II, Civilization, Balance of Power, Falcon 3.0,* and *EF2000*) feature computer-generated opponents that challenge users. Gilman Louie of Spectrum HoloByte estimates that three of the four years required to produce a new video game are dedicated to developing algorithms for controlling computer-generated forces.

Increasingly capable computer-generated opponents have been incorporated into video games since the first commercial video game was introduced in 1970. Nutting and Associates' *Computer Space* allowed users to control a rocket that was pitted against two computer-controlled flying saucers. The player avoided the flying saucer's missiles while trying to steer its missile into one of the saucers. The flying saucers were controlled by a simple random function. After a few short months, players were able to quickly master the game and earn bonus time; soon after, players became bored with the game and stopped playing. Learning from this experience, the designers of the next major video game, *Pong*, replaced computer-generated agents with a second joystick so that players played against each other rather than a computer. In the early 1980s, Atari Games created what many considered to be the first credible strategic military simulation on a personal computer, *Eastern Front*. Players confronted a computer-based challenger that relied on a simple but effective rule-based system. The system used a series of "if-then" statements to determine the computer's response to a player's move. Continued advances in computer technology and computer-generated forces enabled the creation of more sophisticated agents in games such as *Harpoon* and *Aegis,* as well as *Panzer General.* PC-based games further pushed the development of high-quality computer-generated opponents. Unlike most dedicated video games (with game boxes, such as those manufactured by Nintendo, Sega, and Sony), which are designed for simultaneous play by two to four players, PCs are generally used by one person at a time and games are designed for individual play. PC games therefore demand development of effective computer-generated forces, and PCs typically have microprocessors with sufficient power to support more complex computer-generated forces.

Future trends in video games will heighten the need for computer-generated forces and characters. Many multiplayer and on-line games will feature *persistent universes* that continue to exist and evolve even when a particular player is not engaged in the game. These games will take place over a significant span of time, at times without a definitive end. They will be similar to multiple-user domains in that users can come in to the game and exit at will. Persistent universes are critical to effective on-line games because they negate the need to coordinate large number of players and schedule games; in a persistent universe the game is always being played. Such games also get around the need to play an entire campaign or game in one sitting, enabling players to enter and exit as their schedules permit. At the same time, such games pose several problems: (1) the movement of players in and out of the game may be disorienting for the other players and destroys the continuity of the game;[37] (2) game masters will not be able to ensure that enough real players are available at any given time to make the game enjoyable for participants; and (3) it may be hard to generate an environment that gives players a large enough role and provides the necessary rewards to keep them coming back. The solution to some of these problems is to create an environment that is fully populated by computer-controlled forces: either automated forces that are entirely controlled by the computer or semiautomated forces that are given high-level instructions by a real player but are then controlled by the computer. When players enter the environment, they will be able to replace one of the automated elements until they log off; then the computer will regain control. Players may also give general orders before logging off to keep units on a strategic or tactical direction until they return. The entertainment industry has begun to make progress in this area. A handful of real-time games, such as *Command and Conquer,* allow the player to control game pieces at a high level, giving the player instructions as to where to move or what action to take but not specifying the details of the process. Additional work will enable the broader use of such techniques.

Location-based Entertainment

Computer-generated character technologies also appear in immersive virtual reality attractions. Walt Disney Company's *Aladdin* attraction, for example, puts the participant into a virtual environment with many simulated individuals performing throughout the virtual world.[38] Characters such as shopkeepers, camel herders, the sultan's guards, and others populate the simulated city of Agrabah. Participants experience this attraction through head-mounted displays and spatialized headphones. They sit on a saddle and hold a section of Persian carpet that serves as their controller and enables them to fly their magic carpet around the city. As they explore the city, they encounter characters who go about their own business but react to the presence of the participant. Behaviors are programmed into these characters and others. All characters react to the guest's presence; some of them have some intelligence designed into them, and they attempt to provide useful information to the guests, based on their current circumstances.

In other entertainment media, books, film, and television, there are extensive examples of intelligent agents in synthetic worlds. In Warner Brother's 1994 film *Disclosure*, an angelic avatar assists the protagonist in searching a virtual reality library representing the computer's file system. In the current television series, *Star Trek Voyager*, the ship's physician is a holographic projection with intelligence and a database compiled from the expert knowledge of thousands of other physicians.

One of the standards of persistent virtual worlds and characters is the 1992 book *Snowcrash* by Neal Stephenson. *Snowcrash* defines a persistent virtual world known as the Metaverse that is mostly populated by real people who go there by donning an avatar that represents them in that space. There are also totally synthetic characters, of greater or lesser capability and complexity, who interact with real characters in the Metaverse as if they were simply avatars for real people. Today, these complex human-mimicking synthetic characters are simply science fiction. The challenge is to make them a reality for worlds of entertainment as well as worlds for training.

Film

The film industry has also expressed interest in computer-generated characters and digital actors of various kinds. Digital effects studios are now creating three-dimensional digital data sets of actors, such as Tom

Cruise, Denzel Washington, and Sylvester Stallone.[39] Work in digital actors is attempting to generate digitized versions of real actors that can be used in making films. While one motivation for developing digital actors is to avoid the high costs of hiring real actors, the true motivations include the ability to create additional scenes, if necessary, after a real actor has finished a film and to perform actions that real actors cannot or will not do, such as dangerous stunts.

Digital actor technology can also allow directors to create new characters that combine elements of existing actors—or to resurrect dead stars. GTE Interactive, for example, recently unveiled a digitized likeness of Marilyn Monroe—created by Nadia Magnenat-Thalmann—that can chat with visitors on the World Wide Web and respond to typed questions with speech and facial expressions.[40] A short film starring the computer-generated Marilyn, *Rendezvous à Montreal*, also has been created. Digital actors are seen as a key element of interactive media.[41] Interest has become large enough to have spawned a conference on virtual humans in June 1996; a second conference was held in June 1997. The goal of such efforts is not just to incorporate existing two-dimensional video images into film (e.g., as was done in *Forrest Gump*) but to allow films and other sorts of entertainment to be based around digital characters that act and perform various roles.

DOD Applications of Computer-generated Characters

DOD has a strong interest in what it calls computer-generated forces (CGFs). CGFs fall into two categories: (1) semiautomated forces (SAFs) that require some direct human involvement to make tactical decisions and to control the activities of the aggregated force and (2) automated forces, which are completely controlled by the computer. Both kinds of computer-generated forces are under development now for military systems and will find extensive application in modeling and simulation as the technology continues to mature.

DOD systems use CGFs for a variety of applications. In systems designed to train individual war fighters, autonomous forces are used to create adversaries for trainees to engage in simulated battles. In large networked training simulations, computer-generated characters are widely used to control opposing forces since it is too expensive to have all the opposing forces controlled by experts in foreign force doctrine. For such scenarios the military has also pioneered the development of SAFs, which are aggregated forces (such as tank platoons, army brigades, or fighter wings, as opposed to individual tanks, soldiers, and aircraft) that require human control at a high level of abstraction. Rather than controlling specific actions of individual elements of the unit, the SAF operator provides strategic direction, such as moving the unit across the river or flying close air support. The computer directs the forces to carry out the command. Hence, SAFs provide a way for training high-ranking commanders who must control the actions of thousands of soldiers on the battlefield. To date, most training tools have been directed toward stand-alone training simulators and part-task trainers that train individuals and systems such as SIMNET that train small groups of soldiers to work together on the battlefield.

There is now a diverse and active interest throughout the DOD modeling and simulation community in the development of computer-generated forces. DARPA is sponsoring the development of modular semiautomated forces for the Synthetic Theater of War program, which includes both intelligent forces and command forces. This effort also involves development of a command-and-control simulation interface language. It is designed for communications between and among simulated command entities, small units, and virtual platforms. The military services, specifically the Army's Close Combat Tactical Trainer program, are now developing opposing forces and blue forces (friendly forces) to be completed in 1997. The British Ministry of Defence also is developing similar capabilities using command agent technology in a program called Command Agent Support for Unit Movement Facility. DOD has several programs to improve the realism of its automated forces, such as DARPA's Intelligent, Imaginative, Innovative, Interactive What If Simulation System for Advanced Research and Development (I^4WISSARD). Academic and industrial interest in this technology led to the First International Conference on Autonomous Agents in February 1997 in Marina del Rey, California.

Common Research Challenges

The challenge for today's researchers is to develop computer-generated characters that model human behavior in activities such as flying a fighter aircraft, driving a tank, or commanding a battalion such that participants cannot tell the difference between a human-controlled force and a computer-controlled force.[42] Doing so can help prevent participants from looking for and taking advantage of the gaps in a logic routine instead of developing skills that can be applied with or against other human participants (an old pilot's adage is, "if you ain't cheating, you ain't trying!"). In his 1996 chess match with the IBM computer *Deep Blue*, for example, Gary Kasparov learned the computer's tendencies during his first losing game and exploited those tendencies—and the computer's lack of adaptability—to come back and win the tournament. He was less successful in his 1997 match because *Deep Blue* had been programmed to develop strategies like a human and could conceive of and execute moves unanticipated by Kasparov.[43]

The key is to find ways to implement computer-generated characters that behave convincingly like human participants for extended periods of time. Doing so requires research to develop agents that (1) can adapt their behaviors in response to changes in the behavior of human participants, (2) accurately model the behavior of individual entities (as opposed to aggregated units), and (3) can be easily aggregated and disaggregated.

Adaptability

A significant problem facing the development of automated forces is that humans learn and adapt faster than most existing computer algorithms. To date, games have been short enough that computer-generated forces based on specified scripts or simple "if-then" rules could provide enough of a challenge for most players, but with persistent universes simple rule-based systems will not be good enough to control automated forces. SAFs used in DOD systems are typically based on aggregated behaviors of tank or aircraft crews. Operations are extremely regimented; tank elements, for instance, operate on the basis of the Army's combat instruction sets, which were relatively easy to codify. DOD is also attempting to create automated SAFs with behaviors that can adapt. Efforts in these areas are resulting in a shift from SAFs that are algorithm based to ones based on artificial intelligence (AI). The first experiment was with a system called SOAR, developed by Carnegie Mellon University, the University of Southern California, and the University of Michigan.

Most such work is concentrated on knowledge acquisition by the SAF using AI techniques such as expert systems and case-based reasoning. Expert systems are developed by interviewing experts to discover the rules of thumb used to solve problems and then putting that information into the software in a way the computer can use. Case-based reasoning represents an alternative approach in which SAFs learn from their own successes and failures in simulations. The SAF acquires knowledge from new experiences in real time and adds it to its knowledge base for later use. The knowledge base consists of many cases that the computer uses in real time to match the scenario it is facing and respond appropriately. This technique is the basis for machine learning in DOD's I^{4}WISSARD program. Currently, it is done manually during an after-action review, but ideally the computer would do it automatically. For example, a pilot could train the computer by dogfighting with it. If the pilot performed a maneuver the computer had never seen, it would add the trick to its knowledge base and respond appropriately. This could provide a richer training experience for the pilot and make the simulator training useful longer.[44]

Additional experimentation is under way using complex adaptive system techniques to generate new behaviors. As part of a DARPA program, McDonnell Douglas has developed a system that uses genetic algorithms[45] to develop new behaviors and tactics for military simulations. In previous trials, tactics developed by the system were used in simulators and shown to be effective for military operations.

Modeling Individual Behaviors

A related challenge is developing computer-generated characters that mimic *individual* behavior rather than group behavior.[46] Group behavior can often be modeled using statistical modeling and rule-based decision processes. The goal is to develop group actions that seem *reasonable*, but reasonable is an easier test than *human-like*. Did that automated battalion move its tanks in a reasonable and believable fashion? Did the group react using appropriate doctrine? It is easier to hide errors in decision making with a large group of units than with a computer-generated individual. It is more difficult to model and create automated individuals that users can communicate and react to that is believably human. Within DOD, SAF-level simulation has typically gone as low as an entity like a tank or aircraft. Individual soldiers or crew members have not generally been represented, though work is ongoing in this area. As DOD explores more opportunities for training individual soldiers (dismounted infantry), the need to have realistic, intelligent simulated individuals becomes acute.

Human Representations

Research is also needed to develop ways of creating realistic digitized humans that look, move, and express emotion like their real counterparts. DOD has increased its efforts in simulating and modeling individual soldiers in synthetic environments through its Individual Combatant Simulation objective. A joint effort is under way between the U.S. Army's Simulation, Training, and Instrumentation Command (STRICOM) and the Army Research Laboratory. One of the key advances required for developing low-cost solutions to this problem are technologies for visualization of human articulation in real-time networked environments. The entertainment industry is building on motion-capture techniques pioneered by DOD for use in developing games and creating special effects. These techniques track the movements of various joints and extremities as real characters perform a given set of tasks. The data can then be used to create more realistic synthetic effects.

Supporting natural language voice communications also is important. During networked play, voice support is critical for coordinated activities. The computer-generated character must interpret the voice input, react, and acknowledge the user using an equally natural voice. The voice must also indicate stress and the emotional state of the computer-generated character. The user must care about the automated forces and arrive at the same conclusions whether or not the forces under control are human or computer generated. There are many situations in which a player would move a unit into a suicidal situation with a computer-generated character but would choose not to if the force was human.

Aggregation and Disaggregation

Additional research is also needed in the area of aggregation and disaggregation. In large simulations, inactive individual units, such as tanks or soldiers, are often aggregated into higher-level units, such as tank battalions or platoons, to minimize the number of elements the system must track and to allow higher-level control of operations (such as by a field commander). In doing so, information about the individual elements is lost, as only average values of mobility or capability are retained for the aggregated unit. Thus, when the aggregated unit becomes active, the individual elements cannot be disaggregated into their original form. Each tank, for example, will be assigned the average mobility and firepower capabilities of the entire grouping rather than capabilities consistent with those it had before aggregation. Such inconsistencies not only limit the fidelity of the simulation but also generate incongruities among the representations of a simulation perceived by different participants (such as the field commander who sees aggregated levels of capability and tank commanders who see their own disaggregated capabilities). Though not crucial in all simulations or all engagements, such inaccuracies do limit the fidelity of simulations and will become more significant as simulations move toward incorporating individual warriors and participants. Improved methods of aggregation and disaggregation that preserve more state information (possibly in a standardized format) could minimize the amount of information the simulation must retain while preserving greater fidelity and consistency.

Spectator Roles

Another area in which DOD and the entertainment industry have overlapping interests is in developing technology for incorporating spectators into models and simulations. As Jacquelyn Ford Morie noted during the workshop, not everyone involved in digital forms of entertainment will want to be direct participants. Some will prefer to engage as a spectator, similar to sports such as baseball, football, and tennis in which only a small percentage of the participants actually play in a match and much of the industry is built around the fans. Morie believes that "there is a potentially huge market to be developed for providing a substantial and rewarding spectator experience in the digital entertainment realm" (see position paper by Morie in Appendix D). As Morie notes, being a spectator does not necessarily mean being passive; it is about being a participant with anonymity in a crowd, providing a less threatening forum in which people can express themselves.

DOD has already expressed an interest in this type of capability. The role of the "stealth vehicles" has become increasingly important in defense simulations. Such vehicles are essentially passive devices that allow observers to navigate in virtual environments, attach to objects in the environments, and view simulated events from the vantage point of the participant. As multiplayer games become more sophisticated and interesting, such a capability may evolve into a spectator facility that will allow novices to observe and learn from master practitioners. Popular games may evolve to the level of current professional sports with teams, stars, schedules, commentators, and spectators.

TOOLS FOR CREATING SIMULATED ENVIRONMENTS

Another area in which DOD and the entertainment industry have common interests is in the development of software and hardware tools for creating simulated environments. Such tools are used to create and manipulate databases containing information about virtual environments and the objects in them, allowing different types of objects to be placed in a virtual environment and layers of surface textures, lighting, and shading to be added. For games this may be a 3D world that is realistic (such as a flight simulator) or fantastic (like a space adventure), in which an individual interacts directly with the synthetic world and its characters. For film and television, simulated models are often used as primary or secondary elements of scenes that involve real actors, while in other cases the entire story is built around synthetic characters, be they traditional two-dimensional (2D) animations or more advanced 3D animations. For

DOD these worlds are synthetic representations of the battle space (ground, sea, and air) and virtual representations of military systems.

Sophisticated hardware and software tools for efficiently constructing large complex environments are lacking in both the defense and entertainment industries. At the workshop Jack Thorpe of SAIC stated that existing toolsets are quirky and primitive and require substantial training to master, often prohibiting the designer from including all of the attributes desired in a simulated environment (see position paper by Thorpe in Appendix D). Improved tools would help reduce the time and cost of creating simulations by automating some of the tasks that are still done by hand. Alex Seiden, of Industrial Light and Magic, claims that software tools are the single largest area in which attention should be focused. Animators and technical directors for films face daunting challenges as shots become more complicated and new real-time production techniques are developed to model, animate, and render synthetic 3D environments for film and video.

Entertainment Applications and Interests

For digital film and television, special effects and animation are performed during the *preproduction* and *postproduction* processes. Preproduction brings together many different disciplines, from visual design to story boarding, modeling to choreography, and even complete storyboard simulation using 2D and 3D animations. Postproduction takes place after all of the content has been created or captured (live or otherwise) and uses 2D and 3D computer graphics techniques for painting, compositing, and editing. Painting enables an editor to clean up frames of the film or video by removing undesirable elements (such as deleting a microphone and props that were unintentionally left in the scene or an aircraft that flew across the sky) or enhancing existing elements. Compositing systems enable artists to seamlessly combine multiple separate elements, such as 3D models, animations, and effects and digitized live-action images into a single consistent world. Matched lighting and motion of computer graphics imagery (CGI) are critical if these digital effects are to be convincing.

In the games world the needs for content-creation tools are similar. Real-time 3D games demand that real-world imagery, such as photographic texture maps, be combined quickly and easily with 3D models to create the virtual worlds in which pilots fly. In the highly competitive market that computer game companies face, time to market and product quality are major factors (along with quality of game play) in the success of new games. This challenge has been eased somewhat in the past few years as companies have begun offering predefined 3D models and textures that serve as the raw materials that game and production designers can incorporate into their content.

Despite the enormous cost savings that can be enjoyed from automating these processes, entertainment companies invest little in the development of modeling and simulation tools. Most systems are purchased directly from vendors.[47] Film production companies using digital techniques and technologies tend to write special-purpose software for each production and then attempt to recycle these tools and applications in their next production. Typically, little time or funding is available for exploring truly innovative technologies. The time lines for productions are short, so long-term investments are rare. Leveraging commercial modeling and animation tools from both the entertainment world (Alias I Wavefront, Softimage, etc.) and DOD simulation (Multigen, Coryphaeus, Paradigm Simulation) is starting to form a bridge between the entertainment industry and DOD.

DOD Applications and Interests

DOD faces an even greater challenge in its modeling and simulation efforts. Because of the large number of participants in defense simulations, the department requires larger virtual environments than the entertainment industry and ones in which users can wander at their own volition (as opposed to traditional filmmaking in which designers need to create only those pieces of geometry and texture that will be seen in the final film). Beyond training simulations, content-creation tools are potentially useful in creating simulations of proposed military systems to support acquisition decisions. DOD could use such models to prototype aircraft, ships, radios, and other military systems. The key would be linking conceptual designs, computer-aided engineering diagrams, analysis models, or training representations into a networked environment that would enable DOD to perform "what if?" analyses of new products. Finding some way to allow these varied types of data to fit into a common data model would greatly facilitate this process.

Like the entertainment industry, DOD lacks affordable production tools to update simulation environments and composite numerous CGI elements. While its compositing techniques are useful and efficient for developing certain types of simulation environments, they cannot handle the complexity demanded by some high-fidelity applications. Some models and simulation terrain must be built and integrated using motion, scale, and other perceptual cues. Here, DOD personnel encounter problems similar to those of entertainment companies that set up, integrate, and alter CGI environments. Human operators can be assisted by appropriate interactive software tools for accomplishing these iterative tasks.

Having better tools to integrate and create realistic environments could play a major role in the overall simulation design of training systems, exploring simulation data, and updating simulation terrain. Interactive tools could empower more individuals to participate in this process and would increase strategic military readiness.

Research Challenges

Database Generation and Manipulation

Both the entertainment industry and DOD have a strong interest in developing better tools for the construction, manipulation, and compositing of large databases of information describing the geography, features, and textures of virtual environments. Simulations of aircraft and other vehicles, for example, require hundreds or thousands of terrain databases; filmmakers often need to combine computer-generated images with live-action film to create special effects. Most existing systems for modeling and computer-aided design cannot handle the gigabyte and terabyte data sets needed to construct large virtual worlds. As Internet games companies begin to develop persistent virtual worlds and architectural, planning, and military organizations develop more complete and accurate models of urban environments, the need for software that can create and manipulate large graphics data sets will becoming more acute. At DOD the data used to create these databases are typically captured in real time from a satellite and must be integrated into a completed database in less than 72 hours to allow rapid mission planning and rehearsal.

Today's modeling tools can be very powerful, allowing users to create real-time models with texture maps and multiple levels of detail using simple menus and icons. Some have higher-level tools for creating large complex features, such as roadways and bridges, using simple parameters and intelligent modeling aids. At the assembly level, new tools use virtual reality technology in the modeling stage to help assemble large complex environments more quickly and intuitively. Still, modeling tools have not gotten to the point of massive automation. There are some automated functions, but overall responsibility for feature extraction, creation, and simplification is in the hands of the modeler. More research is needed in this area.[48]

Bill Jepson from UCLA is exploring systems for rapidly creating and manipulating large geo-specific databases for urban planning. With a multidisciplinary research team, he has designed a system capable of modeling 4,000 square miles of the Los Angeles region. It uses a client-server architecture in which several multiterabyte databases are stored on a multiprocessor system with a server. Communications between client and server occur via asynchronous transfer mode, at about 6 megabytes per second. Actual 3D data are sent to the client based on the location of the observer, incorporating projections of the observer's motion. Additional research is under way to link this system with data from the Global Positioning System so that the motions of particular vehicles, such as city buses, can be tracked and transmitted to interested parties. Similar systems could be useful for the Secret Service or the Federal Bureau of Investigation for security planning or for U.S. Special Forces or dismounted infantry training operations in a specific geographic locale. Other work at the University of California, Berkeley, is exploring the automatic extraction of 3D data from 2D images.[49] These methods are likely to play a large role in the future in the rapid development of realistic 3D databases.

Another area of possible interest to both the entertainment industry and DOD is in the development of technologies that allow image sequence clips to be stored in a database. This would permit users in both the defense and entertainment communities to rapidly store and retrieve video footage for use in modeling and simulation. A prototype system has been developed by Cinebase, a small company working with Warner Brothers Imaging Technology. Additional development is required to make the technology more robust and widely deployable.

Additional efforts to develop more standardized formats for storing the information contained in 3D simulated environments would be beneficial to both DOD and the entertainment industry. A standard format could be developed that allows behaviors, textures, sounds, and some forms of code to be stored with an object in a persistent database. Such efforts could build on the evolving VRML standard. The goal is to devise a common method for preserving and sharing the information inherent in 3D scenes prior to rendering.[50]

Compositing

Both DOD and the entertainment industry are interested in software tools that will facilitate the process of combining (or compositing) visual images from different sources. Such tools must support hierarchy and building at multiple levels of detail: they must allow a user to shape hills, mountains, lakes, rivers, and roads as well as place small items, such as individual mailboxes, and paint words on individual signs. They must also allow designers to develop simulated environments in pieces that can be seamlessly linked together into a single universe. This need will become more acute as the scale of distributed simulations grows. Existing computer-aided design tools do not have the ability to easily add environmental features, such as rain, dust, wind, storm clouds, and lightning, to a simulated scene.

There are many unsolved compositing problems in pre- and postproduction work for filmmaking that are directly related to simulation and modeling challenges. For example, a need exists for postproduced light models for digital scenes and environments. To create appropriate lighting for composited realistic live-action scenes, lighting models must affect digitized images that were captured under variable lighting conditions. Such a simulation problem is encountered when realistic photographic data are composited into simulation data and the lighting must be interactively adjusted from daylight to night during persistent simulations. Here, it is necessary to develop lighting models that image-process photographic data to provide postproduced lighting adjustments after scenes have been captured. Solutions to these problems do not exist, yet the research would be applicable to both the entertainment industry and DOD.

Opportunities may exist for DOD and the entertainment industry to share some of the advances they have made in designing systems for creating models and simulation. DOD might be able to use some of the advanced compositing techniques that have been developed by the entertainment industry to integrate live-action video with computer graphics models. The entertainment industry's software techniques for matching motion and seamlessly integrating simulated scenes into a virtual environment might also be beneficial to DOD. However, most entertainment software is extremely proprietary. It will be necessary to address proprietary issues and methods of information exchange before extensive collaboration can occur between the entertainment industry and DOD. Conversely, some DOD technologies might prove to be very beneficial for entertainment applications as well. At the workshop, Dell Lunceford, of DARPA, suggested that some of the technologies developed as part of DOD's Modular Semiautonomous Forces (ModSAF) program might be useful in creating some of the line drawings used in preproduction stages of filmmaking. ModSAF cannot support the detailed graphical animation needed for facial expressions, but it could facilitate the simpler earlier stages of production in which characters are outlined and a story's flow is tested.

Interactive Tools

Interactive tools that facilitate the creation of simulations and models and that can be used for real data exploration could be valuable to both the entertainment industry and DOD. The computer mouse and keyboard are extremely limited when creating CGI scenes, and individuals are often impaired or constrained by these traditional input devices. A recent project of the National Center for Supercomputing Applications located at the University of Illinois at Urbana-Champaign resulted in an interactive virtual reality interface to control the computer graphics camera in 3D simulation space. The project created an alternative virtual reality computer system, the Virtual Director, to enhance human operator control and to capture, edit, and record camera motion in real time through high-bandwidth simulation data for film and video recording. This interactive software was used to create the camera choreography of large astrophysical simulation data sets for special effects in the IMAX movie, *Cosmic Voyage*. This project has proven to be valuable for film production as well as scientific visualization. Such uses of alternative input devices to explore and document very large data sets are nonexistent in commercial production because of the time line required to develop such technology, yet this type of tool is extremely important to solve many problems in the entertainment industry as well as DOD simulation and modeling.

CONCLUSION

As this chapter illustrates, the defense modeling and simulation community and the entertainment industry have common interests in a number of underlying technologies ranging from computer-generated characters to hardware to immersive interfaces. Enabling the two communities to better leverage their comparative strengths and capabilities will require that many obstacles be overcome. Traditionally, the two communities have tended to operate independently of one another, developing their own end systems and supporting technologies. Moreover, each community has developed its own modes of operation and must respond to a different set of incentives. Finding ways to overcome these barriers will present challenges on a par with the research challenges identified in this chapter.

NOTES

1. For a more comprehensive review of research requirements for virtual reality, see National Research Council. 1995. *Virtual Reality: Scientific and Technological Challenges*, Nathaniel I. Durlach and Anne S. Mavor, eds. National Academy Press, Washington, D.C.

2. DOD has several ongoing programs to extend the military's command, control, communications, computing, intelligence, surveillance, and reconnaissance systems to the dismounted combatant. These include the Defense Advanced Research Projects Agency's Small Unit Operations Program, Sea Dragon, Force XXI, and Army After Next.

3. Latency is not the only factor that causes simulator sickness, and even completely

eliminating latency will not eliminate simulator sickness. See position paper by Eugenia M. Kolasinski in Appendix D.

4. This subsection is derived from a position paper prepared for this project by the Defense Modeling and Simulation Office; see Appendix D.

5. Sheridan, T.B. 1992. *Telerobotics, Automation, and Human Supervisory Control.* MIT Press, Cambridge, Mass.

6. For a more complete description of the SIMNET program see Van Atta, Richard, et al., 1991, *DARPA Technical Accomplishments, Volume II: An Historical Review of Selected DARPA Projects,* Institute for Defense Analyses, Alexandria, Va., Chapter 16; and U.S. Congress, Office of Technology Assessment, 1995, *Distributed Interactive Simulation of Combat,* OTA-BP-ISS-151. U.S. Government Printing Office, Washington, D.C., September.

7. U.S. Congress, Office of Technology Assessment, *Distributed Interactive Simulation of Combat,* p. 32, note 6 above.

8. Gilman Louie, Spectrum Holobyte Inc., personal communication, June 19, 1996.

9. Pausch, Randy, et al. 1996. "Disney's Aladdin: First Steps Toward Storytelling in Virtual Reality," *ACM SIGGRAPH '96 Conference Procedings: Computer Graphics.* Association for Computing Machinery, New York, August.

10. RTime Inc. introduced an Internet-based game system in April 1997 that supports 100 simultaneous players and spectators. See *RTIME News*,Vol. 1, February 1, 1997.

11. The National Research Council's Computer Science and Telecommunications Board has another project under way to examine the extent to which DOD may be able to make better use of commercial technologies for wireless untethered communications. A final report is expected in fall 1997. Another project to examine DOD command, control, communications, computing, and intelligence systems was initiated in spring 1997.

12. Specifications for implementing multicast protocols over the Internet are outlined by S.E. Deering in "Host Extensions for IP Multicasting," RFC 1112, August 1, 1989, available on-line at http://globecom.net/ietf/rfc1112.html. See also Braudes, R., and S. Zabele, "Requirements for Multicast Protocols," RFC 1458, May 1993.

13. As such, multicast stands in contrast to *broadcast*, in which one designated source sends information to all members of the receiving community, and to *unicast* systems in which a sender transmits a message to a single recipient.

14. This capability is called *routing spaces.* It will permit objects to establish *publish regions* to indicate areas of influence and *subscription regions* to indicate areas of interest. When publish and subscription regions overlap, the RTI will cause data to flow between the publishers and the subscribers. The goal of this effort, and the larger Data Distribution Management Project, of which it is part, is to reduce network communications by sending data only when and where needed. See Defense Modeling and Simulation Office, *HLA Data Distribution Management: Design Document Version 0.5*, Feb. 10, 1997; available on-line at http://www.dmso.mil/projects/hla/.

15. Internet Engineering Task Force, "Large Scale Multicast Applications (lsma) Charter," available on-line at http://www.ietf.org/html.charters/lsma-charter.html.

16. Much of the material in this section is derived from a position paper prepared for this project by Will Harvey of Sandcastle Inc.; see Appendix D.

17. Deployment of a new algorithm for queue management, called Random Early Detection, may help greatly reduce queuing delays across the Internet.

18. Floyd, S., and V. Jacobson. 1993. "Random Early Detection Gateways for Congestion Avoidance," *IEEE/ACM Transactions on Networking* 1(4):397-413; Wroclawski, J. 1996. "Specification of the Controlled-Load Network Element Service," available on-line as ftp://ftp.ietf.org/internet-drafts/draft-ietf-intserv-ctrl-load-svc-03.txt.

19. Clark, D. 1996. "Adding Service Discrimination to the Internet," *Telecommunications Policy* 20(3):169-181.

20. Sandcastle Inc., an Internet-based game company, is one source of research on synchronization techniques.

21. DOD defines *modeling and simulation interoperability* as the ability of a model or simulation to provide services to and accept services from other models and simulations and to use the services so exchanged to enable them to operate effectively together. See U.S. Department of Defense Directive 5000.59, "DOD Modeling and Simulation (M&S) Management," January 4, 1994, and U.S. Department of Defense, Under Secretary of Defense for Acquisition and Technology, *Modeling and Simulation (M&S) Master Plan*, DOD 5000.59-P, October 1995.

22. All participants in a simulation do not need an identical representation of the environment. Individual combatants, for example, will differ from fighter pilots in the amount of terrain they can see and the sensor data (radar, infrared, etc.) available to them. The key is ensuring that their views of the environment are consistent with one another (e.g., that all players would agree that a given line of trees obstructs the line of sight between two participants in the simulation).

23. DIS conveys simulation state and event information via approximately 29 PDUs. Four of these PDUs describe interactions between entities such as tanks and personnel carriers; the remainder transmit information on supporting actions, electronic emanations, and simulation control. The *entity state* PDU is used to communicate information about a vehicle's current position, orientation, velocity, and appearance. The *fire* PDU contains data on weapons or ordinance that are fired or dropped. The *detonation* PDU is sent when a munition detonates or an entity crashes. The *collision* PDU is sent when two entities physically collide. The structure of each PDU is regimented and changed only after testing and subsequent discussion at the biannual DIS workshops convened by the Institute for Simulation and Training at the University of Central Florida.

24. Macedonia, Michael R. 1995. "A Network Software Architecture for Large-Scale Virtual Environments." Ph.D. dissertation, Naval Postgraduate School, June; available from the Defense Technical Information Center, Fort Belvoir, Va.

25. Defense Modeling and Simulation Office, *HLA Management Plan: High-Level Architecture for Modeling and Simulation, Version 1.7*, April 1, 1996.

26. The Navy alone has over 1,200 simulation systems that do not currently comply with HLA. A compliance monitoring reporting requirement and waiver process, similar to the Ada waiver process, were put into place. Each affected service is to fund retrofits of simulation systems from their own budgets.

27. Ordering information is available on the DMSO Web site at http://www.dmso.mil.

28. The Computer Science and Telecommunications Board workshop provided an opportunity for representatives from Internet game companies to learn more about HLA. Several agreed to review the specifications to see if they would be applicable to them

29. Lantham, Roy. 1996. "DIS Workshop in Transition to. . . What?," *Real Time Graphics* 5(4):4-5.

30. National Research Council. 1995. *Virtual Reality: Scientific and Technological Challenges*, Nathaniel I. Durlach and Anne S. Mavor, eds. National Academy Press, Washington, D.C.

31. Macedonia, Michael R., et al. 1995. "Exploiting Reality with Multicast Groups," *IEEE Computer Graphics & Applications*, September, pp. 38-45.

32. Brutzman, Don, Michael Zyda, and Michael Macedonia. 1996. "Cyberspace Backbone (CBone) Design Rationale," paper 96-15-99 in *Proceedings of the 15th Workshop on Standards for DIS*, Institute for Simulation and Training, Orlando, Florida; Brutzman, Don, Michael Zyda, Kent Watsen, and Michael Macedonia. 1997. "Virtual Reality Transfer Protocol (vrtp) Design Rationale," accepted for the *Proceedings of the IEEE Sixth International*

Workshop on Enabling Technologies: Infrastructure for Collaborative Enterprises (WETICE '97), held June 18-20, 1997, at the Massachusetts Institute of Technology, Cambridge, Mass.

33. Brutzman et al., 1996, "Cyberspace Backbone (CBone) Design Rationale," and Brutzman et al., 1997, "Virtual Reality Transfer Protocol (vrtp) Design Rationale," note 32 above.

34. Macedonia, "Exploiting Reality with Multicast Groups," note 31 above.

35. A standard 14.4-kilobit-per-second modem can transmit or receive a standard DIS packet in approximately 80 milliseconds, meaning that only about five players can participate in a real-time interactive game if each must send and receive messages (updating positions, velocities, etc.) to and from each other player at each stage in the game and latencies must be kept below 100 milliseconds.

36. From this perspective the code base is like the way a television studio thinks of it sets, props, and sound stages. The code base needs to be able to be data driven so that new episodes can be created in less than a week instead of a couple of years. Programming will be developed using scripting tools that allow writers and designers to quickly develop new stories. These tools will be important to help the writers and designers not only create new environments but also direct automated units and characters to "perform" new roles for the new scenarios.

37. For example, a player may be flying an F-15 along with a wingman when a pair of enemy MiGs engages them in battle. As a player breaks into a turn, he or she may realize that the wingman has disconnected (intentionally or unintentionally) from the game.

38. The *Aladdin* attraction is something of an anomaly in that Walt Disney Imagineering approached it not only as a theme park attraction but also as scholarship. It published results of its research in the open literature. See Pausch, Randy, et al. 1996. "Disney's Aladdin: First Steps Towards Storytelling in Virtual Reality," *ACM SIGGRAPH '96 Conference Proceedings: Computer Graphics*. Association of Computing Machinery, New York, pp. 193-203.

39. Fryer, Bronwyn, "Hollywood Goes Digital," available on-line at http://zeppo.cnet.com/content/Features/Dlife/index.html.

40. Ditlea, Steve. 1996. "'Virtual Humans' Raise Legal Issues and Primal Fears," *New York Times*, June 19; available on-line at http://www.nytimes.com/library/cyber/week/0619humanoid.html.

41. Magneanat Thalmann, N., and D. Thalmann. 1995. "Digital Actors for Interactive Television," *Proceedings of the IEEE*, August.

42. An agent that could meet this requirement would satisfy the "Turing test." Alan Turing, a British mathematician and computer scientist, proposed a simple test to measure the ability of computers to display intelligent behavior. A user carries on an extended computer-based interaction (such as a discussion) with two unidentified respondents—one a human and the other a computer. If the user cannot distinguish between the human and the computer responses, the computer is declared to have passed the Turing test and to display intelligent behavior.

43. Chandrasekaran, Rajiv. 1997. "For Chess World, A Deep Blue Sunday: Computer Crushes Kasparov in Final Game," *Washington Post*, May 12, p. A1.

44. U.S. Congress, Office of Technology Assessment. 1995. *Distributed Interactive Simulation of Combat*, OTA-BP-ISS-151. U.S. Government Printing Office, Washington, D.C., September, pp. 123-125.

45. Genetic algorithms are computer programs that evolve over time in a process that mimics biological evolution. They can evolve new computer programs through processes analogous to mutation, cross-fertilization, and natural selection. See Holland, John H. 1992. "Genetic Algorithms," *Scientific American*, July, pp. 66-72.

46. The National Research Council is conducting another project on the representation of human behaviors in military simulations. See National Research Council. 1997. *Repre-*

senting Human Behavior in Military Simulations—Interim Report, Richard W. Pew and Anne S. Mavor, eds. National Academy Press, Washington, D.C.

47. Paul Lypaczewski of Alias|Wavefront estimates that the market for off-the-shelf modeling and simulation tools is about $500 million per year.

48. See National Research Council, *Virtual Reality,* note 30 above.

49. Debevec, P.E., C.J. Taylor, and J. Malik. 1996. "Modeling and Rendering Architecture from Photographs: A Hybrid Geometry- and Image-based Approach," *Proceedings of SIGGRAPH '96: Computer Graphics.* Association of Computing Machinery, New York, pp. 11-20.

50. The process of rendering computer graphics is the process of making frames from objects with motion so they can be displayed by the computer or image generator.

3

Setting the Process in Motion

Encouraging the defense and entertainment industries to work together to strengthen the technology base for modeling and simulation will require more than a common research agenda. As workshop participants noted, large cultural barriers exist between the U.S. Department of Defense (DOD) and the entertainment industry that will impede attempts to work together. If collaboration is to succeed, these barriers must be overcome and a process must be established to facilitate the kinds of collaboration that could advance this nation's modeling and simulation technology base. Consideration needs to be given to coordinating the research agendas of the two communities, such as through greater sharing of information, and to structuring cooperative research. In addition, both DOD and the entertainment industry will need to work—individually and collectively—to ensure the continued viability of the research base from which both communities draw. They must ensure that human resources are developed that have the requisite skills for creating effective computer-based simulations and that adequate funding is provided to sustain the research base, particularly at the university level. Doing so will require additional input and advice from the academic community, whose participation in the workshop was limited.

OVERCOMING CULTURAL BARRIERS

The entertainment industry and DOD are two different cultures, with different languages, different business models, and separate communities of constituents (Box 3.1). To date, they have operated relatively independently of one another. Though some sharing of technology and research has occurred, much of the technology transfer has been mediated through the university research community; few direct connections have been made between companies actively engaged in developing entertainment products and services and DOD.

Movement of people between DOD and the entertainment industry is limited. Many people from the entertainment industry were reluctant to participate in the workshop—or serve on the steering committee that convened the workshop—because of the project's connection with defense and their impressions that few opportunities exist for collaboration. Nevertheless, workshop participants provided personal examples of the key movements of people between the two communities that have cross-pollinated each other's efforts: Eric Haseltine, vice-president of research and development and chief scientist at Walt Disney Imagineering, began his career in flight simulation at Hughes Aircraft Company; and Carl Norman, a senior producer with the games company Strategic Simulations Inc., is a former Marine officer who later worked on simulation and training systems for the Corps. Yet most workshop participants agreed that a movement of people between defense and entertainment is not the trend. Jordan Weisman, of Virtual World Entertainment, remarked that employee migration between the entertainment industry and DOD is minimal and that this is a contributing factor to the minimal amount of technology transfer between the two communities. Such differences both influence and are influenced by differences in the business models that the two communities follow. Overcoming them will require efforts to improve communication between members of the two communities.

Different Business Models

DOD and the entertainment industry differ significantly in their goals, motivations, and methods of doing business. These differences make it difficult for the communities to work together to advance the technology base for modeling and simulation, but with sufficient interest on both sides, ways can be found to overcome these obstacles. While few formal attempts at coordinating research or conducting joint research have been tried, a handful of companies have successfully transitioned from defense work to commercial work, demonstrating the possibility of success. Further attempts to facilitate greater coordination of research activities between DOD and the entertainment industry will have to build upon these examples to find ways to bridge the gap between the two communities.

BOX 3.1
From DOD to Entertainment: A Personal Journey

Robert Jacobs, president and director of Illusion Inc., recounted at the workshop some of the cultural differences he has noticed between the defense modeling and simulation community and the entertainment industry. Illusion Inc. has its roots in defense simulation, owing to work on DOD's simulation network (SIMNET) project. In 1994 the company merged with another company as part of an effort to develop simulation-based systems for entertainment. As of 1996, roughly 80 percent of Illusion's revenues derived from entertainment systems.

> What we discovered immediately was an enormous culture shock. Our entertainment partners had a different way of thinking than we did. The basis for pricing defense work, for example, is cost. There are several different kinds of contractual formats that are used to govern what defense contractors can charge for their time and for their products, but all of them have as the base what it costs to do the job and some allowance for reasonable profit—the amount of which depends on whether the client or the contractor takes the risk. In the commercial world, things are very different. You get a good idea, you invest your own money and effort in bringing that idea to fruition, and then you charge what the market will bear. What the market will bear depends on what the other companies are doing and what the perceived value of the product is to the ultimate consumer, the public. You have to be pretty confident of your instincts in that kind of an arena to invest significant money in bringing a product to market. If you don't make it, you can't go back for an overhead rate adjustment and you can't tell the client "Gee, I'm sorry I underestimated this thing, and you've got to help me out here."
>
> I suggested that Illusion's new partners ought to fill out time sheets and allocate their time to the jobs they were working on. They didn't understand why that was necessary because they said, "We're going to work all year, and at the end of the year what's left over after we pay everybody is the profit. So why do we have to keep track of all this stuff?" Well, we keep track of all that stuff because we still do DOD work, too, and the auditors want to see what we're doing all the time, not just while we're working on defense work.
>
> There is also a big difference in what is considered important in the business. We had a client come in and ask us to describe the technical approach for the system we were developing. I proudly launched into a very detailed description of the marvelous new motion system that we were using on a ride we were building and how it has advanced servo actuators, and it is very fast and high performance, and it was a terrific bargain compared to what was available in the marketplace. I noticed that the client's eyes were glazed over; what the client really wanted to know was what the ride was like and what the story was that accompanied it. So, I let my partner from the entertainment industry describe it, and the client was much happier. I thought that the technical details were terrific, but nobody cares about that kind of thing. What they want to know is what they're going to get for the money they pay in terms of the customer's experience.

Some of the more notable differences between the DOD and entertainment communities are listed below:

- *Time horizons.* Military research and development programs typically span several years; it is not unusual for a large-scale program to take a decade or more to complete. Products are often designed to be used for long periods of time, with requisite upgrades and support.[1] Entertainment projects are generally much shorter in duration. Game companies may spend up to three years developing a new product; film companies will also spend a year or two producing a new film. Little attention is given to upgrades and support for older products. Many game companies, in fact, intend for their games to become obsolete soon so that customers will buy new versions.
- *Market structure.* Defense contractors serve a single customer (DOD) that, though comprised of many heterogeneous parts, operates with a fairly consistent business model and a standard set of procurement policies. Typically, DOD specifies up front the requirements of a system it wants developed and invites contractors to design a system that will meet the specifications. Contractors are generally reimbursed for their expenses plus a fixed profit. Entertainment companies, in contrast, serve diverse markets, with varied tastes and business models. Companies must identify market needs, formulate creative concepts that addresses those needs, and invest their own capital in developing the concept. They must also invest considerable effort in marketing new products to consumers, who may or may not buy the product in the end.
- *Profitability.* Most defense contracts specify the profit a contractor may make on a given project, typically on the order of 6 percent. Entertainment companies have no guarantees on profits and assume large risks with each undertaking. At the same time, successful projects in the entertainment industry can generate profit margins of 100 percent or more; the film industry relies on a limited number of such blockbuster hits to underwrite its operating expenses and compensate for the more numerous failures. Nevertheless, the great potential profitability of entertainment products limits the enthusiasm with which entertainment companies would embrace defense-related work.
- *Research.* DOD operates a large research and development program with fairly centralized budgeting and direction. Military needs are translated into technical requirements that spur research programs. In 1996, DOD spent $688 million on basic and applied research in computer science and mathematics. The entertainment industry lacks any coordinated direction of research. Individual firms select research projects that meet their business needs. According to Alex Seiden, of Industrial Light and Magic, most companies in the entertainment industry do not con-

duct long-term research; scientists and engineers in film and game production environments, in particular, are primarily interested in solving the problems at hand on a short time line coupled to a movie premier or game release—often only three months. A few entertainment companies, such as Disney, do invest more in longer-term research and development products, but such companies are few in number.

- *Intellectual property.* Entertainment companies have strong concerns over intellectual property rights and endure great pains to ensure control. Despite the rapid product cycles in the entertainment industry, many products build on technology incorporated into earlier products or adhere to proprietary standards and architectures that have longer useful lifetimes. While defense contractors have similar interests and concerns, working for the federal government often implies a lack of control over intellectual property rights that entertainment companies would not tolerate.

Such differences pose hurdles to any attempts to bring the entertainment industry and DOD closer together to develop modeling and simulation technologies. Indeed, defense contractors have generally failed in their attempts to diversify into commercial markets of many kinds because of the great differences in business practices.[2] Defense contractors who spend internal research and development money on projects with commercial implications find they have great difficulty spinning off successful projects to commercial industry.[3]

Similarly, the entertainment industry complains that DOD's bureaucracy inhibits its desire to do defense-related work. Many small companies, such as video game designers and manufacturers of peripheral devices, lack the resources (money and staff) to adhere to DOD's acquisition regulations, which cover both the purchase of finished products and many research and development activities. Given the growing markets for entertainment products and the potential profitability of successful products, many entertainment companies do not find working with DOD to be an attractive alternative to their usual entertainment efforts. According to Buzz Hoffman, of ThrustMaster Inc., "developing products for the mass commercial market offers far more opportunities for profit than Defense Department contracts." Until the potential benefits of collaboration become more apparent to the entertainment industry and the associated costs can be reduced, formal collaboration will be difficult to initiate.[4]

Facilitating Coordination and Cooperation

A first step toward bridging the gap between the defense modeling and simulation community and the entertainment industry would be to encourage greater sharing of information between the two communities.

DOD and the entertainment industry have a great deal to learn from one another, and open collaboration and cross-fertilization could allow each to tap into the other's knowledge and better target its own research programs. Eric Haseltine, of Walt Disney Imagineering, claimed at the workshop that "the thing that the entertainment industry can get the most from DOD is just knowing what's been done, so they don't have to reinvent the wheel. If there were some places where [the entertainment industry] could go to find out what's been done before we go off and do it ourselves, that would be hugely valuable."

Few participants from DOD or the entertainment industry seem to know how to effectively exchange information between the two communities, nor are they generally aware of the mechanisms that might exist. Many workshop attendees agreed that connections between DOD have been informal and based on a limited number of personal contacts; there is a general frustration in both communities regarding ways to connect to solve problems that might be common to both. A number of workshop attendees complained that it sometimes seems easier to get information for product development from outside sources rather than government agencies, precluding many opportunities for collaboration. At the same time, workshop participants acknowledged that the entertainment industry is very fragmented and that no centralized mechanism exists for collecting information or learning about the needs of the industry as a whole. While some attempts have been made, through such organizations as the Technology Council of the Society of Motion Picture and Television Engineers, they have been limited in scope and success. Because the entertainment industry is very competitive and proprietary, entertainment companies do not freely share information about research programs or interests, and the frequent changes in employment make individual connections fleeting.

One mechanism for promoting informal information sharing between members of the defense modeling and simulation community and the entertainment industry and breaking down cultural barriers is to encourage attendance at relevant conferences that cross industry boundaries. To date, entertainment companies have not been well represented at conferences on defense modeling and simulation, nor have members of the defense industry been prominent at entertainment industry conferences. The Association of Computing Machinery's annual Special Interest Group in Computer Graphics (SIGGRAPH) conference provides a forum and marketplace for exciting ideas combining research, art, education, and business in computer graphics but lacks a significant military presence. Other important conferences are the Computer Game Developer's Conference, which has seminars and roundtable discussions on many topics relevant to modeling and simulation, and the annual Electronic Entertainment Expo, which has large exhibits and presentations related to interactive entertainment. Several members of the entertainment industry stated that DOD's lack of participation at such conferences is a major factor behind their belief that the department has little to offer the entertainment industry in the way of relevant research or technology. By presenting more work at such conferences or setting up booths on convention floors, DOD could help artists, software developers, and executives in the entertainment industry learn about relevant DOD technology and research. At the same time, DOD would be able to see and hear about current developments in the entertainment industry that will enter the marketplace 18 to 24 months later and start to make the social connections that always facilitate information flow.

One way to help bridge the gap between DOD and entertainment industry standards initiatives is for representatives from each community to attend the other's meetings. For example, the High-level Architecture (HLA) community's Architecture Management Group (AMG) could incorporate representatives of the Internet community and of the networked games community, thereby providing them a platform for discussions and information exchanges. Greater representation from the distributed interactive simulation (DIS) community also might be considered by the AMG. In return, the communities engaged in developing Virtual Reality Modeling Language (VRML) and virtual reality transfer protocol (*vrtp*) standards could encourage DOD's HLA community to attend its forums, such as the VRML Consortium. Currently, the only DOD representation at that consortium is through a membership paid for by the NPSNET Research Group of the Naval Postgraduate School. The Defense Modeling and Simulation Office (DMSO) should consider joining the VRML Consortium to benefit from the work going on there and from the contacts.

While seats on each other's technical guidance boards are one method of collaboration, another is establishment of an annual workshop similar to the successful DIS workshops. The workshop could build on the initial effort hosted by the Computer Science and Telecommunications Board (CSTB) as part of the present project, which demonstrated the mutual benefits of bringing together members of the defense and entertainment communities. The goal of an annual workshop would be to attract technical papers and presentations from DOD, the entertainment industry, and the Internet community. Unlike the DIS workshop, the papers for this new workshop would be peer reviewed and the workshop should be held at a university center set up to shepherd the process and to provide a neutral meeting ground for defense and entertainment participants, similar to the Institute for Simulation and Training at the University of Central Florida. The workshop could be funded jointly by industry and DOD.

Additional progress toward collaboration in interoperability standards may emerge from the newly formed Simulation Interoperability Standards Organization (SISO). SISO is a reorganization and replacement of the group that developed DIS standards. It will continue to expand the semiannual workshops started by DIS but will separate standards development work from the workshops. SISO was established primarily to broaden interests from the real-time, platform-level simulations of DIS to all forms of interoperable simulations. SISO's mission is to (1) continue to support DIS standards, (2) assume responsibility for commercializing DOD's HLA standards, and (3) develop simulation interoperability standards for any organization that needs them. SISO sponsored a one-day seminar in January 1997 and its first Simulation Interoperability Workshop in March 1997, both of which tried to attract participation from the entertainment industry. To date, participation by the entertainment industry appears to have been limited by a lack of interest in interoperability standards and by the perception that SISO's interest is in defense simulation and propagation of the HLA.

At CSTB's workshop, several different models were discussed for facilitating collaborative work that could benefit both defense and entertainment. One possibility is for joint funding of university research in technologies associated with modeling and simulation, such as computer graphics, networking, and computer-generated characters. University research has played a key role in developing modeling and simulation technology and in disseminating that technology, largely through students, to both the entertainment and the defense industries. Alternatively, a few companies have demonstrated the possibility of operating in both the defense and the entertainment communities, either by developing technologies and products that support both industries or simply drawing on research in both fields to make separate defense and entertainment products.[5] As Robert Jacobs, of Illusion Inc., stated, "We want to stay in the defense world as well as the entertainment world because of the access to wonderful technology that we recognize exists there. We think that the technology transfer happens much more effectively if the company is working in both environments."

The federal government has also established programs to encourage transfers of technology and cooperative research between government and commercial organizations. The Small Business Innovative Research program, which earmarks a percentage of federal research and development budgets for small business, requires award winners to submit plans for commercializing their technologies. The Federal Technology Transfer Act requires all federal laboratories with research and development budgets above a certain threshold to each establish an Office of Research and Technology Applications to promote technology transfer. These offices are authorized to both license federal technology to commercial industry and to enter into cooperative research and development agreements

(CRADAs) with commercial companies. Under CRADAs, federal laboratories and private companies may work jointly on projects of mutual interest: laboratories may contribute researchers or facilities to an endeavor; industry may contribute researchers, facilities, or funding. In general, laboratories are authorized to negotiate intellectual property agreements that reflect the relative contributions of government and industry to a project.[6] In the defense modeling and simulation community the Technology Transfer Program of the Naval Air Warfare Center Training Systems Division is playing a lead role. The Navy's Office of Training Technology and DMSO's Information Analysis Center also coordinate technology transfer activities, but it is not clear that such activities have successfully transferred technology to the commercial sector.

Two programs exist within the federal government to foster linkages between government laboratories and industry. The Federal Laboratory Consortium helps industry understand technology transfer and assists with referrals to appropriate government agencies. The National Technology Transfer Center, a federally funded center located in West Virginia, also helps researchers from industry solve problems by linking them with relevant experts at federal laboratories. Such programs are not well publicized or well known within the entertainment industry and therefore have not served as effective vehicles for sharing information as yet.

Other techniques might be useful in stimulating collaborative work between DOD and the entertainment industry. At the workshop, Jordan Weisman suggested funding competitions to stimulate the entertainment industry to work on topics that are interesting to DOD researchers. Such competitions do not need to have large monetary rewards, just great promotional opportunities. For example, a $10,000 prize has been established for computer chess systems. Alternatively, an Internet site could be set up specifically to exchange information between DOD and the entertainment industry concerning hot topics in modeling and simulation. Often there is no need to transfer algorithms; just knowing that someone else has solved a problem and seeing the solution in action will often spur the creation of better solutions. Thus, dissemination of information on existing developments might be enough to enable the two communities to benefit from each other's successes.

HUMAN RESOURCES

Workshop participants agreed that one of the most significant steps the entertainment industry and DOD can take to jointly advance modeling and simulation is to ensure a continued supply of well-educated workers and researchers. Both the entertainment industry and DOD are critically dependent on skilled workers who understand how to develop simulations that are visually pleasing and who can resolve the technical problems associated with large-scale distributed simulations, such as latency, graceful degradation of performance as scale increases and network latencies lengthen, and maintenance of consistent state information across large numbers of simulators. The development of workers with a mixture of technical and artistic capabilities represents a particular challenge because of its interdisciplinary nature. Whereas computer science and electrical engineering departments will train technical workers to address questions about networking and distributed simulation, the creation of visually literate workers demands cooperation between engineering and art departments, which are separated by large cultural and institutional gaps.

Demand for workers with an understanding of the artistic and technical considerations embodied in modeling and simulation is growing faster than the supply.[7] Such people are important not only in entertainment and defense but in manufacturing industries as well, where they can help design automobile, aircraft, and defense systems. Workshop participants indicated a shortage of talented, high-quality, experienced people to develop virtual environments, modeling and simulation software, digital animation, design, and scripting of virtual worlds. "Ask the production manager of any effects studio," reported Alex Seiden, of Industrial Light and Magic, "and [he or she] will tell you their biggest problem is the shortage of skilled animators and technical directors." Several workshop participants noted that the rapid pace at which technology is evolving is reducing the number of individuals who know how to effectively exploit the new technologies. Scott Watson, of Walt Disney Imagineering, suggested that experienced programmers who deal with multiplayer multiprocessor technical problems are at a premium. He estimated that less than 5 percent of the programming population understands such issues.

As Ed Catmull, of Pixar Animation Studios, noted, U.S. entertainment companies are raiding foreign countries for talented animators. As a result, salaries for talented animators are rising rapidly. Starting salaries for animators range from $60,000 to $100,000 per year. Production companies tend to bid up salaries as they hire away each other's workers in order to learn about new technologies and techniques. Continued salary growth in the entertainment industry could make DOD less able to attract top talent. It has already experienced problems retaining top information technology officers, both because of limited avenues for advancement at DOD and lucrative offers from commercial industry.[8] According to John Latta, a consultant with 4th Wave Inc., rapid expansion of three-dimensional graphics capabilities (in hardware and software)

> **BOX 3.2**
> **Visual Literacy**
>
> Tom West, author of *In the Mind's Eye,* discusses visual literacy in relation to advances in computer graphics: "It will be left to humans to maximize what is most valued among human capabilities and what machines cannot do—and increasingly these are likely to involve the insightful and integrative capacities associated with visual modes of thought." Three-dimensional (3D) computer graphics experts extrapolate from software algorithms to graphical models and scenes; visual literacy informs individuals with creative solutions that help bridge this gulf and provide innovation. For example, an extremely valuable skill is efficiently enhancing a 3D simulated environment with realistic light, shade, and geometry while successfully managing computational and technical constraints. Mastering the economy of detail, color, compositing, and visualization techniques can provide realism to simulations. Knowledge of these types of computer graphics techniques involves a combination of visual literacy and technical expertise. While this highly developed combination of skills is difficult to find in a single individual, workshop attendees argued that it would provide more powerful simulations that bring about believability and convincing virtual worlds, with computational economy. This type of skill set directly overlaps with the needs of both the entertainment industry and the DOD simulation community. Visual literacy crosses into the boundary of art and design education, and many workshop attendees believe it important that such educational needs be reported.
>
> ---
>
> SOURCE: West, Thomas G. 1991. *In the Mind's Eye: Visual Thinkers, Gifted People with Learning Difficulties, Computer Images, and the Ironies of Creativity.* Prometheus Books, Buffalo, New York, p. 254.

will continue to strain the talent pool, especially among women and minorities who are already underrepresented in this type of work. According to many workshop participants from the entertainment industry, the shortage of qualified workers is the single most confounding issue in the digital film effects and gaming industries today.[9]

Digital artists and designers are particularly valuable in creating visual models and designing graphical interface tools. Eric Haseltine, of Walt Disney Imagineering, has found that the best-designed computer user interfaces are not created by computer scientists or human factors engineers but by artists and designers who are visually literate (Box 3.2). While Disney funds and employs many university graduates with such training, the demand is much greater than the supply. Likewise, digital designers are in demand for creating virtual spaces for the Internet. Microsoft Corporation is employing increasing numbers of designers for its *VChat* and other social virtual spaces. The company has found that interdisciplinary and multigender teams of artists and computer scientists create the richest interactive social spaces for the Internet.[10]

The need for visually literate workers may also become more apparent in the defense industry, as DOD attempts to make its simulations more engaging and believable to participants. Fidelity will continue to be important in DOD's training simulations, but the fidelity of defense simulations is only part of the greater problem. As Danny Hillis, of Disney, observed at the workshop:

> If you think of the basic problem of what these [training] simulators are for, it is not a problem of *simulation*. We are not trying to simulate the reality of, let's say, a tank. The problem is not to get something that looks like a tank. The problem is instead to cause a change in the person's mind, so that when they get into a real tank in a battlefield, they do the right thing.

In this view, the goal of a simulation is not to approximate reality as nearly as possible, but to present individuals with the appropriate set of cues to produce the training effect desired. Creating the desired change in a person's mind requires a suspension of disbelief in the individual who is experiencing, interacting, and making decisions in the simulation. It requires a complex combination of attributes that engage and teach the user. Being able to express such traits in virtual environments is a communication skill, and animators are trained in ways to map human behavior to models and motion. Many workshop participants (from both the entertainment industry and DOD) believe that these types of talents have historically been considered less important than technical skill and fidelity in defense training applications. But as one DOD participant noted, "DOD needs people telling them things they never knew about how their systems could be used."

Other nontechnical skills also are becoming more important to modeling and simulation. For example, one challenge is to model human and cultural behavior in realistic interactive virtual spaces. Creating a sense of social space in virtual reality is becoming more important to DOD, according to some workshop participants. Kirstie Bellman, of the Defense Advanced Research Projects Agency, noted that there is a sense of social reality that is important in certain simulations, such as those represented by multiple user domains. Understanding human behavior in these artificial realities involves the expertise of multiple disciplines, combined as teams. Learning why some of these places work while others fails often requires the insight of anthropologists. Increasing demands for DOD to participate in noncombative activities, such as drug interdiction and peacekeeping, demand increasing understanding of the human component of behaviors, which needs to be incorporated into modeling and simulation. While DOD has not historically funded some of these areas, it is important to reexamine the funding domain with which it operates.

Workshop participants from both the entertainment industry and DOD agreed that cross-disciplinary programs blending computer science and art are needed to provide workers with the education to support both entertainment and defense applications of modeling and simulation. Few university programs currently exist that combine visual literacy, digital design, computer science, and engineering; most universities have separate departments—or separate schools—for engineering and the arts. Nevertheless, the Naval Postgraduate School recently established a master's degree program in modeling, virtual environments, and simulation. After several years of operation as a two-year program in Vancouver, the DigiPen School plans to open a four-year accredited program in Seattle, Washington, that will concentrate on training students in the creation of video games. The program combines courses in physics, math, and computer science with art and storytelling.[11] The goal of such programs is not to provide graduates with narrow training in specific technologies or systems used for creating simulated environments but to ensure that they are well grounded in the disciplines involved in modeling and simulation.

Creating such programs presents a challenge. According to Alex Singer, an independent film director, some U.S. universities demonstrate tremendous resistance to the whole range of computer studies within the arts. At the workshop, Gilman Louie, of Spectrum HoloByte Inc., commented that "if you're in a computer science program, the art department won't let you into their advanced classes. The same with computer science departments—unless you change your major." This observation is often a result of classes being oversubscribed in a major's courses that relate to digital imaging; however, the shortage of cross-disciplinary courses is related to the structure of universities. Few U.S. universities provide faculty with incentives to teach across disciplines; universities are divided into departments where funding and promotion are discipline specific. While there has been increasing emphasis on interdisciplinary work at funding agencies, such as the National Science Foundation, most such efforts have tended to focus on scientific and engineering fields, not on merging science and engineering with art and design.

Trying to create an academic discipline that combines science and engineering with art creates specific problems. As with other fields of science and engineering, computer science rewards researchers for publishable research, the results of which are typically expressed mathematically. Fields that combine computer science with art (e.g., computer graphics, virtual environments) have difficulty being recognized as serious areas of study, noted Michael Zyda, of the Naval Postgraduate School, because they are new fields that are often viewed as applications of computer science rather than core research areas. Proponents will have to convince the computer science community that work in this field will yield publishable results and that the arts can play more than a supportive role. According to one reviewer of this report, the SIGGRAPH community has been working on this problem for several years, with limited results.

DOD and the entertainment industry could take a more active role in encouraging the development of such programs, using existing funding mechanisms as a lever. While DOD has not historically funded educational programs in digital arts and design, these areas do appear to have a growing relevance to defense needs. Other government agencies with interests in education and work-force issues also may have an interest in such programs.

To further promote strengthening of the education system in this country, both the entertainment industry and DOD communities could promote internships to help students and administrators gain a better understanding of real-world applications of modeling and simulation. Formal arrangements might be made between universities, the entertainment industry, and DOD research labs to create programs for internships. Such programs may encompass more than just universities. During her tenure at the Visual Systems Laboratory at the Institute for Simulation and Training, Jacquelyn Ford Morie helped create such a program for undergraduate and high school students, the "Toy Scouts." Through this program, students were given access to the laboratory's virtual reality equipment and were given opportunities to develop innovative projects. Researchers and professionals from the entertainment industry, DOD, and the laboratory mentored the students, who developed several innovative applications of virtual reality technology (see the position paper by Morie in Appendix D). Similar programs at the high school and vocational levels might be effective in filling the need for skilled workers in modeling and simulation.

MAINTAINING THE RESEARCH BASE

Closely related to the problem of developing human resources is the need to maintain a strong base for research into modeling and simulation, especially at the university level. DOD has historically played a major role in funding university research in computer science. As recently as 1996, DOD provided nearly half of all federal funding for university research in computer science; the National Science Foundation (NSF) provided most of the rest.[12] Early DOD investments in university research spurred development of many technologies that have turned into billion-dollar industries that now lie at the core of both entertainment and defense modeling and simulation: workstations, graphics technology, and virtual reality.[13] Such funding for university research has not only produced new knowledge and new technologies but also provided research opportunities that are significant component of students' education. "People," says Ed Catmull, "are the best products from research dollars."

Many workshop participants believe that the funding environment has changed in ways that could prove detrimental to the long-term viability of the technology base for modeling and simulation.[14] Attempts to reduce the federal budget deficit and trim defense spending have put additional pressure on federal research expenditures. While total research funding for computer science has continued to grow in real terms, funding for basic research has remained flat since 1990. Most of the growth in research funding is attributable to increases in funding for applied research. Workshop participants claim that government-funded research projects are now more product oriented than they used to be.[15] Research contracts—even for projects considered basic and applied research—often specify particular products, completion dates, and interim deliverables. University researchers claim that this shift, while responsive to calls for greater accountability in publicly funded research, is incompatible with fundamental research, which by its very nature is speculative and unpredictable and more likely than product-oriented research to generate fundamental change. As Ed Catmull noted at the workshop:

> The thing that has benefited us all in the past are those programs where people are free to pursue wilder visions, where [researchers] can't see things exactly clearly, but they themselves become the foundations on which we build growth in the future.

Others noted that NSF funding, though it has grown in recent years, entails a higher degree of administrative overhead; the need to generate frequent proposals has taken critical research time away from researchers, including those at the NSF-funded Science and Technology Center for Computer Graphics and Scientific Visualization.[16]

Industry funding for university research has not compensated for the change in federal funding. Though total industry contributions to university research have grown in recent years, industry still supports just a fraction of university research.[17] Workshop participants claimed that industry support for research in digital arts, virtual environments, and distributed simulation is especially small. Many entertainment companies do not support university research because they are small and see themselves as users of technology more than developers. In addition, industry funding of university research is generally linked to specific industry needs that are closely coupled with business priorities. As Paul Lypaczewski of Alias|Wavefront noted at the workshop, companies are in the business of generating profits, so they fund those schools and those programs whose area of research and competencies are aligned with the types of problems that companies want to fix or understand better. Companies also expect nearer-term returns from their research investments than does the government. While industry-funded university research is not necessarily incorporated immediately into products, it must move technology forward in ways that ultimately benefit industry; thus, funding decisions center around finding bodies of research that are applicable to both industry and university. Corporate sponsors also tend to demand greater control over intellectual property that results from research programs, creating a requirement for greater secrecy, which limits the dissemination of new research knowledge.

In addition, several workshop participants suggested that industry is not contributing equipment to university laboratories as it did in the past. University researchers perceive a decline in contributions by traditional donors and note that many of the newer industry leaders have not stepped up to the challenge to provide donations. The lack of computer hardware and software technology affects the availability and quality of technology for research and training in universities by limiting the kinds of systems on which students may work.

Many of these concerns mirror changes seen in funding for science and technology generally. The demise of high-profile national initiatives to invigorate scientific and technical research programs (such as the space race or the Cold War),[18] constraints on federal budgets for science and engineering, and changes in patterns of industrial research and development (including the restructuring of corporate research laboratories at such companies as IBM, Xerox, and AT&T) influence the nature of research and development in many fields. Resolution may therefore require action beyond the modeling and simulation community. Nevertheless, to the extent that the modeling and simulation community can succinctly and accurately convey the nature of its concerns, such as through this report, it may be able to influence the process.

CONCLUDING REMARKS

As evidenced by the workshop, strong commonalities exist between defense and entertainment applications of modeling and simulation and in the technologies needed to support them. Whereas DOD has traditionally led the field and provided a significant portion of related funding, the entertainment industry has made rapid advances in 3D graphics generation, networked simulation, computer-generated characters, and immersive environments. Aligning the research agendas of these two communties to allow greater coordination of research developments, sharing of information, and collaborative research could provide an opportunity to more rapidly and economically achieve the goals of both the defense and entertainment industries.

Linking these two communities represents a significant challenge; differences in business practices and culture need to be overcome in order to find mechanisms for cooperation and collaboration. Additional efforts will be needed to ensure adequate education of visually literate people who can create engaging simulated environments and to ensure funding for continued research. Success will rely on sustained commitment from both sides—and from a shared belief that the benefits of collaboration are worth the costs. This workshop represented the first step toward exploring the benefits and the costs of such collaboration; the fact that it attracted many participants from the entertainment industry, as well as DOD, suggests that some degree of mutual interest exists. Additional steps will need to be taken to capitalize on that interest.

NOTES

1. DOD has recently expressed some interest in leasing computer hardware and software for its systems—especially training systems—in order to shorten the time required to acquire new technologies.

2. U.S. Congress, Office of Technology Assessment. 1992. *After The Cold War: Living with Lower Defense Spending.* U.S. Government Printing Office, Washington, D.C., Chapter 6.

3. Lunceford, Dell, Defense Advanced Research Projects Agency, personal communication, November 6, 1996.

4. However, as noted in Chapter 1, at least one project is under way to develop a system that can be used both as a military training device and a game, demonstrating that cultural obstacles can potentially be overcome.

5. Bray, Hiawatha. 1997. "Battle for Military Video Game Niche On," *Boston Globe*, April 16, p. 1.

6. For a more complete discussion of federal technology transfer efforts and CRADAs, see U.S. Congress, Office of Technology Assessment, 1993, *Defense Conversion: Redirecting R&D,* U.S. Government Printing Office, Washington, D.C.

7. Armour, Barry. 1997. "A Different Kind of Artist," *Computer Graphics,* February, pp. 23-25.

8. Lardner, Richard. 1997. "The Future of Army Automators," *Inside the Army*, April 29.

9. Such sentiments have been expressed outside the workshop as well. Scott Ross, president of Digital Domain, a leading digital effects studio, also has noted a growing need for technical directors, animators, compositors, and digital artists in general. See "Hollywood Reporter," *New Media,* July 31, 1996.

10. Stone, Linda. 1996. "On-line Multimedia Communities," speech before the Fourth Annual Living Surfaces Conference: Design for the Internet, Chicago, Ill., November 16.

11. Wolkomir, Richard. 1996. "The School Where It's OK to Major in Fun and Games," *Smithsonian*, December, pp. 86-97.

12. In 1996 federal obligations for university research in computer science and mathematics totaled $593 million. DOD provided $280 million, and NSF contributed $261 million. Of the DOD funding, DARPA provided $184 million. See National Science Foundation. 1996. *Federal Funds for Research and Development—Fiscal Years 1994, 1995, 1996.* National Science Foundation, Washington, D.C., Table C-58.

13. Computer Science and Telecommunications Board, National Research Council. 1995. *Evolving the High-Performance Computing and Communications Initiative to Support the Nation's Information Infrastructure.* National Academy Press, Washington, D.C.

14. The discussion of changes in the size, scope, and nature of research and development has implications far beyond those for the modeling and simulation community and are not discussed in detail in this report. CSTB has another study in progress, "Information Technology Research in a Competitive World," that will more fully examine the implications of structural changes in the conduct of research and development in information technology in universities, government, and industry.

15. National Science Foundation. 1996. *National Patterns of R&D Resources: 1996.* National Science Foundation, Washington, D.C., Tables C-25, C-26, and C-27. Conversion to constant 1996 dollars is based on gross domestic product deflators contained in Table C-1.

16. The center includes computer graphics programs at Brown University, the California Institute of Technology, Cornell University, the University of North Carolina at Chapel Hill, and the University of Utah. It has a long-term research mission (11 years) to help improve the scientific bases for the next generation of computer graphics environments (both hardware and software). Its research is directed toward modeling, rendering, interaction, and performance.

17. In real terms, total industry funding for university research (in all fields of science and engineering) grew nearly 60 percent between 1987 and 1996, but it still represents just 7 percent of total university research funds. See National Science Foundation. 1996. *National Patterns of R&D Resources: 1996.* National Science Foundation, Washington, D.C., Table C-2.

18. The space program generated many technologies related to modeling and simulation. Satellite imagery and mapping together have been a significant driver of imaging technology.

Appendixes

A

Workshop Agenda

Saturday, October 19, 1996

9:00 a.m.	**Welcome and Introductory Remarks**
	Michael Zyda, Committee Chair
	Anita Jones, Director of Defense Research and Engineering
	Danny Hillis, Walt Disney Imagineering
	Ed Catmull, Pixar Animation Studios
10:15	**Break**
10:30	**Session 1: Electronic Storytelling**
	Moderator: Donna Cox
	Panelists: Alexander Singer, Alex Seiden, Rebecca Allen
12:15 p.m.	**Lunch,** Dining Terrace
1:30	**Session 2: Strategy and War Games**
	Moderator: Gilman Louie
	Panelists: Kirstie Bellman, Peter Bonanni, Paul Chatelier, Mat Genser, Carl Norman, Scott Randolph
3:00	**Break**
3:15	**Session 3: Experiential Computing and Virtual Reality**
	Moderator: Joshua Larson-Mogal
	Panelists: Ben Delaney, Mark Bolas, Scott Watson, Bill Jepson, Eugenia Kolasinski, Traci Jones
5:15	**Reception,** Dining Terrace
6:00	**Dinner,** Atrium

Sunday, October 20, 1996

9:00 a.m.	**Session 4: Networked Simulation**
	Moderator: Michael Zyda
	Panelists: Warren Katz, Robert Jacobs, Brian Blau, Will Harvey, David King
11:00	**Break**
11:15	**Session 5: Low-Cost Simulation Hardware**
	Moderator: Jordan Weisman
	Panelists: David Clark, Buzz Hoffman, Jeff Potter, Gary Tarolli
12:15 p.m.	**Lunch**, Refectory
1:30	**Session 6: Finding Common Ground**
	Moderator: Paul Lypaczewski
	Panelists: Jim Foran, John Latta, Jacquelyn Ford Morie, Jack Thorpe
3:00	**Synthesis: Setting the Research Agenda**
	Moderator: Michael Zyda
	Panelists: Steering Committee
3:45	**Adjourn**

B

Workshop Participants

Rebecca Allen
University of California at
　Los Angeles

Kirstie Bellman
Defense Advanced Research
　Projects Agency

Brian Blau
Intervista Software

Mark Bolas
Fakespace Inc.

Peter Bonanni
Virginia Air National Guard

F. T. Case
Defense Advanced Research
　Projects Agency

Edwin E. Catmull
Pixar Animation Studios

Paul Chatelier
Defense Modeling and Simulation
　Office

David Chen
RGB Technology Inc.

David Clark
Intel Corporation

Ronald Cobb
Artist

Judith Dahmann
Defense Modeling and Simulation
　Office

Ben Delaney
CyberEdge Information Services
　Inc.

Clark Dodsworth, Jr.
Osage Associates

James Foran
Silicon Graphics Inc.

John Geddes
Ames Research Center

Mathias Genser
Spectrum HoloByte Inc.

Martin Gundersen
University of Southern California

Will Harvey
Sandcastle Inc.

Eric Haseltine
Walt Disney Imagineering

Janet Weisenford Healy
Naval Air Warfare Center

W. Daniel Hillis
The Walt Disney Company

Buzz Hoffman
ThrustMaster Inc.

James Hollenbach
Defense Modeling and Simulation Office

Robert Jacobs
Illusion Inc.

Mark Jefferson
Defense Modeling and Simulation Office

Bill Jepson
University of California at Los Angeles

Anita Jones
Office of the Secretary of Defense

Traci Jones
U.S. Army Simulation, Training & Instrumentation Command

David King
Total Entertainment Network

Eugenia Kolasinski
Consultant

John Latta
4th Wave Inc.

Richard Lindheim
Paramount Television Group

George Lukes
Defense Advanced Research Projects Agency

Dell Lunceford
Defense Advanced Research Projects Agency

Farid Mamaghani
Consultant

Dennis McBride
Office of Naval Research

William McQuay
USAF Wright Laboratory

Jacquelyn Ford Morie
Walt Disney Feature Animation Studios

Carl Norman
Strategic Simulations Inc.

Jeffrey Potter
REAL 3D

David Pratt
Joint Simulation Systems/Joint
  Program Office

Scott Randolph
Spectrum HoloByte Inc.

Vance Saunders
Ball Aerospace and Technologies
  Corporation

Alex Seiden
Industrial Light and Magic

Steven Seidensticker
Consultant

Sandeep Singhal
IBM T.J. Watson Research Center

Gary Tarolli
3Dfx Interactive

Jack Thorpe
Science Applications International
  Corporation

Scott Watson
Walt Disney Imagineering

C

Biographical Sketches of Committee Members

MICHAEL ZYDA is a professor in the Department of Computer Science at the Naval Postgraduate School (NPS) in Monterey, California. Dr. Zyda is also an academic associate and associate chair for academic affairs in that department. He has been at NPS since February 1984. Dr. Zyda's main research focus is computer graphics, specifically the development of large-scale, networked, three-dimensional virtual environments and visual simulation systems. Dr. Zyda was a member of the National Research Council committee that produced the report *Virtual Reality: Scientific and Technical Challenges*. He is also the senior editor for virtual environments for the MIT Press quarterly *PRESENCE*, the journal of teleoperations and virtual environments. Dr. Zyda has been active with the Symposium on Interactive 3D Graphics and was the chair of the 1990 conference, held in Snowbird, Utah, and chair of the 1995 symposium, held in Monterey, California. Dr. Zyda began his career in computer graphics in 1973 as part of an undergraduate research group, the Senses Bureau, at the University of California, San Diego. He received a B.A. in bioengineering from the University of California, San Diego, in 1976; an M.S. in computer science/neurocybernetics from the University of Massachusetts, Amherst, in 1978; and a D.Sc. in computer science from Washington University, St. Louis, Missouri, in 1984.

DONNA J. COX is a professor in the School of Art & Design and associate director for technologies in the School of Art at the University of Illinois, Urbana-Champaign. She is also codirector for Scientific Communications and Media Systems at the National Center for Supercomputing Applications. Ms. Cox has exhibited computer images and animations in more than 100 invitational and juried exhibits during the past nine years, including shows at the Bronx Museum of Art in New York, the Everson Art Museum in New York, the Feature Gallery in Chicago, the Feature Gallery in New York City, the Fermilab in Chicago, and the Museum of Contemporary Photography in Chicago. She has authored many juried papers on computer graphics and scientific visualization and received the Coler-Maxwell Medal for Excellence in 1989. Her work has been reviewed or cited in more than 75 publications, including *Time*, *National Geographic*, *Wall Street Journal*, and *IEEE Computer Graphics and Applications*. Ms. Cox spent a sabbatical working on an IMAX film, *Cosmic Voyage*, for the Smithsonian Institution's National Air and Space Museum. As associate producer for scientific visualization and art director, she has orchestrated scientific visualization software, data, and design for Pixar Animation Studios, Santa Barbara Studios, Princeton University, the University of California at Santa Cruz, the San Diego Supercomputer Center, and the National Center for Supercomputing Applications. Ms. Cox received an M.F.A. in computer graphic arts and a B.A. from from the University of Wisconsin-Madison.

WARREN J. KATZ is vice president and cofounder of MäK Technologies. His responsibilities include corporate operations, new business development, and program management. MäK's corporate goal is to provide cutting-edge research and development services to the Department of Defense in the areas of distributed interactive simulation (DIS) and networked virtual reality (VR) systems and to convert the results of this research into commercial products for the entertainment and industrial markets. MäK's first commercial product, the VR-Link™ developer's toolkit, is the most widely used commercial DIS interface in the world. It is an application programmer's toolkit that makes possible networking of distributed simulations and VR systems. The toolkit complies with the Defense Department's DIS protocol, enabling multiple participants to interact in real time via low-bandwidth network connections. VR-Link is designed for easy integration with existing and new simulations, VR systems, and games. From June 1987 to October 1990, Mr. Katz worked for Bolt, Beranek, and Newman on the SIMNET project. He was the resident drive-train simulation expert, responsible for mathematical modeling of the physical systems and software development. Mr. Katz received B.S. degrees in mechanical engineering and electrical engineering from the Massachusetts Institute of Technology.

JOSHUA LARSON-MOGAL is manager of product strategy for Silicon Graphics' Light Client Division. He is responsible for supervising product management of the division's products and for driving innovation into commercial applications for the range of markets addressed by the division. He was previously the manager of the Enabling Technologies Group in the Advanced Systems Division at Silicon Graphics, where he oversaw a group of product managers working on the OpenGL, IRIS Performer, Open Inventor, REACT, and ImageVision Library software products and the real-time and virtual reality market/technology spaces. In previous positions at Silicon Graphics, Mr. Larson-Mogal served as manager of the market development group, market manager for emerging technologies, product manager for advanced graphics systems, and senior graphics software developer. In these positions he identified new growth markets for advanced graphics hardware and initiated Silicon Graphics' participation in markets for visual simulation, virtual reality, and interactive entertainment. He also initiated the product planning process for the Infinite Reality graphics subsystem, the follow-on to Reality Engine, managed the Power Vision (VGX) graphics workstation products, and developed feature-based solid modeling applications for computer-integrated design, analysis, and manufacturing. In 1985 Mr. Larson-Mogal founded Deneb Robotics Inc., where he designed the system architecture and developed the user interface for IGRIP, a software application for robot work-cell simulation and off-line training. As a graphics software developer at Auto-Trol Technology Corporation, he developed device-independent graphics libraries to support both computer-aided design and computer-aided facilities management applications. Mr. Larson-Mogal received a B.S. degree in computer science from Cornell University.

GILMAN LOUIE has been chair of Spectrum HoloByte Inc. since 1992. In 1982 Mr. Louie founded Nexa Corporation, a developer of entertainment software that later merged with Spectrum HoloByte. Mr. Louie is the creator of the best-selling Falcon air combat simulation, one of the company's leading brand franchises. He received a B.S. in business administration from San Francisco State University.

PAUL LYPACZEWSKI is part of the management team continuing to build and manage Alias|Wavefront in Toronto. He is working with the former executive vice-president of Wavefront to form a distributed development organization of engineers and support staff in Toronto, California, Vancouver, Santa Barbara, and Paris. Mr. Lypaczewski continues to manage corporate research and development (R&D) and oversees all levels of development from product planning, product release, and strategic account management for all of the Toronto products. Mr. Lypaczewski joined Alias Research Inc. in February 1992 as part of a management turnaround team. In his role as vice-president of product development, he oversaw the restructuring of R&D and all levels of development from product planning to product release for all Alias products and was involved in legal and intellectual property issues associated with R&D. Prior to joining Alias, Mr. Lypaczewski worked at CAE Electronic Ltd., a producer of real-time systems, including flight training simulators, air traffic control and energy control systems, and space systems, such as the controls for the Space Shuttle's Canadarm. Mr. Lypaczewski joined the company as an autopilot systems engineer and held a variety of management positions, including senior manager of simulator programs engineering and manager of avionics simulation. In these positions he was responsible for all project engineering and sales proposal support for flight simulation and computer-based training systems. Mr. Lypaczewski received a B.Eng. degree from McGill University and is a member of L'Ordre des Ingenieurs du Quebec.

RANDY PAUSCH is an associate professor of computer science, human-computer interaction, and design at Carnegie Mellon University. He received a B.S. in computer science from Brown University in 1982 and a Ph.D. in computer science from Carnegie Mellon in 1988. He has been a National Science Foundation Presidential Young Investigator and a Lilly Foundation Teaching Fellow. In 1995 he spent a sabbatical with the Walt Disney Imagineering virtual reality studio. Dr. Pausch is the author or co-author of five books and more than 50 reviewed journal and conference proceedings articles, is an active consultant with both Walt Disney Imagineering and Xerox PARC, and has served on a number of National Research Council panels.

ALEXANDER SINGER began his career as a photojournalist and educational filmmaker. His 30-year directing career has resulted in 250 television shows, several full-length feature movies, and many commercials. His directorial credits include *Profiles in Courage, Police Story, The Fugitive, Run for Your Life, Hill Street Blues, Lou Grant, Cagney and Lacey, Star Trek: Voyager,* and *Deep Space 9.* Mr. Singer won an Emmy for an episode of *The Bold Ones* (1972) and represented the series *Police Story* (1975) and *Lou Grant* (1979) for their Emmys. He has lectured on film production, cinematography, and directing and has taught courses at private institutions, universities on two continents, the University of California at Los Angeles extension, and for the Directors Guild of America Special Projects. In addition to his directorial work, Mr. Singer has, for the past several years, been a member of the Global Business Network, a consulting group based in San Francisco with wide-ranging concerns centered on the global economy. Recently, Mr. Singer was under contract as a film consultant to Universal Studio's Orlando theme park and MCA/Matsushita. His work at MCA/Matsushita centered on the development of an entertainment application for virtual reality technology.

JORDAN WEISMAN is chief creative officer for Virtual World Entertainment Inc. This title recognizes his pivotal role as the principal creative architect at Virtual World. Acclaimed as one of the world's premier game and software designers, Mr. Weisman has led the company to its present position atop the fledgling "experience" industry. In 1980 Mr. Weisman and his partner, Ross Babcick, formed the FASA Corporation, a fantasy role-playing board-game publishing company. As FASA's president, Jordan codesigned two of the top five best-selling games in the industry, *BattleTech* and *Shadowrun*. FASA now publishes multiple lines of fantasy and science fiction novels based on its game universes. It was at FASA that Mr. Weisman began to develop the principles behind the interactive games that Virtual World Entertainment now practices at Virtual World. Virtual World opened its BattleTech Center in Chicago in August 1990. As the first location-based virtual reality center in the world, it gave the public a taste of a technology that was formerly the private domain of the National Aeronautics and Space Administration and the military. Mr. Weisman has received numerous awards for game design and has lectured extensively on virtual reality and game design around the world.

D

Position Papers

Prior to the Computer Science and Telecommunications Board's October 1996 workshop on modeling and simulation, participants were asked to submit a one- to three-page position paper that responded to three questions:

1. How do you see your particular industry segment evolving over the next decade (i.e., how will markets and products evolve)?
2. What technological advances are necessary to enable the progress outlined in your answer to question 1? What are the primary research challenges?
3. Are you aware of complementary efforts in the entertainment or defense sectors that might be applicable to your interests? If so, please describe them.

This appendix reproduces a number of these position papers. The papers examine technologies of interest to the entertainment industry and the U.S. Department of Defense, as well as some of the barriers to collaboration. Several of the papers are cited in the body of the report; substantial portions of some have also been incorporated there.

BRIAN BLAU
VRML: Future of the Collaborative 3D Internet

INTRODUCTION

VRML (virtual reality modeling language) is the three-dimensional computer graphics interchange file specification that has become the standard for Internet-based simulations. It is being used in many industries, and the momentum of the standard and industry acceptance continues to grow at a fast pace. Most of the major software and hardware corporations are now starting serious efforts to build core VRML technologies directly into business applications, scientific and engineering tools, software development tools, and entertainment applications.

One of the most significant developments in the history of VRML was its adoption by Silicon Graphics Inc. (SGI), Netscape, and Microsoft during 1995-1996. This broad level of industry acceptance continues to challenge the VRML community to provide an official international standard so that wide adoption will be possible. Given that creation of VRML came from a unique and open consensus-based process, its future depends on continued innovation in the directions of true distributed simulations as well as efforts to keep the standards process moving forward.

HISTORICAL DEVELOPMENT OF VRML

Over the past two years the development of a standard for distributing 3D computer graphics and simulations over the Internet has taken the quick path from idea to reality. In 1994 a few San Francisco cyberspace artisans (Mark Pesce, Tony Parisi, and Gavin Bell) combined their efforts to start the VRML effort. Their intention was to create a standard that would enable artists and designers to deliver a new kind of content to the browsable Internet.

In mid-1995 VRML version 1.0 emerged as the first attempt at this standard. After an open Internet vote, VRML 1.0 was to be based on Silicon Graphics' popular Open Inventor technology. VRML was widely evaluated as unique and progressive but still not useable. At this point broad industry support for VRML was coalescing in an effort to kick-start a new industry. Complimentary efforts were also underway to deliver both audio and video over the Internet. The general feeling was that soon the broad acceptance of distributed multimedia on the Internet was a real possibility and that VRML would emerge as the 3D standard.

After completion of the VRML 1.0 standard, the VRML Architecture Group (VAG) was established at SIGGRAPH 1995 and consisted of eight

Internet and 3D simulation experts. In early 1996 VAG issued a request for proposals on the second round of VRML development. The call was answered by six industry leaders. Through an open vote SGI emerged as the winner with its Moving Worlds proposal. By this time over 100 companies had publicly endorsed VRML, and many of them were working on core technologies, browsers, authoring tools, and content. At SIGGRAPH 1996 VAG issued the final VRML 2.0 specification and made a number of other significant announcements.

To help maintain VRML as a standard, VAG made several concrete moves. First, it started the process of creating the VRML Consortium, a nonprofit organization devoted to VRML standard development, conformance, and education. Second, VAG announced that the International Standards Organization (ISO) would adopt VRML and the consensus-based standardization process as its starting place for an international 3D metafile format.

DISTRIBUTED AND MULTIUSER SIMULATIONS USING VRML

Based on the current state of technology, it is now obvious that distributed 3D simulations are clearly possible for a wide audience. Distributed simulation is the broad term that defines 3D applications that communicate by standards-based communications protocols. Military training, collaborative design, and multiuser chat are examples of such applications.

Widespread adoption of this technology depends on the following key technology factors: platforms, rendering, multimedia, and connectivity. Today, the most popular platforms for accessing the Internet are desktop machines—namely, Windows 95/NT and the Macintosh PowerPC family. These operating systems are running on computing platforms powerful enough to display complex 3D-rendered scenes. The tools are readily available as well, thanks to Microsoft's DirectX media integration API's and ActiveX Internet controls as well as Netscape's Live3D and LiveConnect developer platforms. These software tools, combined with powerful desktop processors, make it easy for software developers to create VRML technologies and products.

Another key aspect of development is the tight integration of multimedia into these platforms. Hollywood and the video games industry see the desktop PC as the next major platform for delivery of multimedia content. This means VRML technology development will be accessible to developers of all types of integrated Internet-based media.

The final key is development of open-protocol communications standards suited for Internet use. Currently, the military uses distributed interactive simulation (DIS) as the communications protocol for training applications and has been successful to date. The integration of DIS with Internet technology is key but not the entire solution. DIS was developed only for military applications. Its broader acceptance by industry is dependent on significant changes to its infrastructure, including the simulation model, numerical representation, integration with VRML, and dependence on Department of Defense initiatives.

Another complementary area of interest is multiuser VRML spaces. These applications are the next step in on-line human-to-human communication and are enabled by the Internet and VRML. Several companies have products that let individuals directly interact with others. In these on-line worlds each person views a fully interactive 3D VRML world, including moving graphical avatars that are the virtual representations of their human counterparts. Some of these applications also include real-time voice that is syncopated with movements of the avatar's eyes and mouth. It is very compelling to communicate with someone and only be able to see their virtual representation.

Several companies and organizations are now starting to collaborate on a standard for VRML-based avatars. These groups are now in the formative stages and are being published by fairly small companies. The first avatar standard will roll out later in 1996.

FUTURE DIRECTIONS

VRML technology and content development in 1996-1997 will focus on several areas. On the standards front, the VRML Consortium and ISO will continue to broaden acceptance of VRML. The VRML Consortium will have its first official meetings in late 1996. Creating the organization and filling it with technical, creative, process-oriented people will be a goal. The VAG will continue to serve as the focus for standards-based VRML work until the consortium is self-sustaining. Also during 1997, ISO will officially adopt VRML as the only international 3D metafile format for the Internet. Once the VRML Consortium is operational, the focus of activities will be on continued development of the VRML specification and the creation of working groups.

On the software and hardware development fronts many advances will be made. VRML 2.0 browsers will emerge and will integrate directly into the popular HTML-based browsers. Manufacturers of three-dimensional hardware accelerators will add features that directly support basic VRML graphics. Tool manufacturers, such as polygonal modelers and scene creation tools, will incorporate VRML read-and-write capabilities. Integration of DIS and other distributed simulation communications protocols will quickly help content authors build multiuser capabilities into their worlds. Finally, content developers will enjoy the flood of new modeling and programming tools.

Given all of these advances there are still three immediate technical areas that need to be addressed before VRML becomes widely adopted: common scripting language, external API, and binary file format. Currently, these areas are quite controversial, but it is clear within the VRML community that solutions to the problems are within reach.

VRML RESOURCES ON THE INTERNET

http://vag.vrml.org—Official home of the VRML spec and the VAG
http://sdsc.vrml.org—Very comprehensive list of VRML resources
http://www.intervista.co—Popular VRML browser
http://www.microsoft.com/ie/ie3/vrml.htm—Popular VRML browser
http://www.sgi.com/cosmo—Popular VRML browser
http://home.netscape.com/eng/live3d—Popular VRML browser
http://www.blacksun.com—Multiuser 3D application
http://www.onlive.com—Multiuser application with real-time voice
http://www.dimensionx.com—Java-based VRML tools
http://www.ktx.com—VRML tools

MARK BOLAS

INTRODUCTION

If the National Aeronautics and Space Administration's VIEW laboratory marks the beginning of the virtual reality (VR) industry, the industry is just about to pass its 10-year mark. There is a rule of thumb stating that it takes about 20 years for a new technology to find its way into the mainstream economy. Applied here, this means 10 years before VR is in the mainstream economy. This prediction seems completely reasonable, or even pessimistic. Consumers can currently purchase VR headsets with integrated tracking for less than $800. A handful of automotive manufacturers and aerospace contractors use VR on an ongoing basis to solve design and engineering problems. However, early adopters are incorporating the technology into their work and lives. They face all of the frustrations and challenges typically associated with being on the cutting edge. The next 10 years will see the VR industry evolve in a straightforward and boring fashion—early adopters will have paved the way for easy use by the mainstream.

This evolution will require a fundamental shift in the way VR technology is viewed and used. The technology must cease to stand apart; it needs to become an invisible part of a user's lifestyle and work habits. This requires progress on two basic fronts: First, the technology must be integrated into the user's physical environment. Second, it must be integrated into the user's software environment.

EVOLUTION

For mainstream users to benefit from VR technologies, the technologies must become pervasive. They must extend throughout our industries and lives. They must diligently work for their users and quietly become part of their lifestyle. The facsimile machine is an example of a technology that has accomplished this.

Walkmen, dishwashers, televisions—All these have become pervasive by thoroughly changing the way people do things. A person does not talk about using a walkman, or a dishwasher, or a television. If anything, a person discusses the content or end result as opposed to the

---

NOTE: The industry segment described here is defined as industries that benefit from immersive human-computer interfaces. The term virtual reality is intended to include this definition.

actual device. "I heard a good song," "The dishes are clean," "Did you see that stupid show last night?"

There is little question that three-dimensional (3D) graphics and simulation are on the way to becoming pervasive. In industry the design process is being transformed to demand 3D models and simulations. This Christmas consumers will be choosing between the Sony or Nintendo platforms with 3D graphics capability being assumed.

However, the VR industry must evolve to provide such 3D systems with immersive interfaces that multiply the utility and effect of the 3D graphics. Currently, most 3D graphics are shown on a 2D screen and manipulated via a 2D mouse. These interfaces effectively remove much of the value present in the 3D environments. The VR industry must maintain the utility and comfort present in a user's natural ability to perceive and manipulate 3D environments and objects.

ADVANCES

For VR to become a pervasive tool, it must become integrated into both the user's physical and software environments. Seamless integration with a user's physical environment is not simple because immersive interfaces tend to immerse—that is, they surround and envelop the user. This can easily intrude on a user's physical and mental environment. The VR industry needs to minimize this intrusion to the point where immersive interfaces are as natural to use as a telephone or mouse. It is interesting to note that both these examples are not inherently natural, but both have been integrated into users' workspaces and lifestyles.

To achieve a natural interface, paradigms that transcend the standard goggles-and-gloves paradigm need to be pursued. The fact that people collaborate, multitask, and eat while they work are down-to-earth aspects that must be considered in the design of immersive tools.

Equally challenging is the integration of these new interfaces in the software environment. Application software packages have typically been written for 2D screens and interfaces. As a result, most immersive interfaces are poor retrofits onto existing packages that were never designed to incorporate them. This lack of integration severely cripples the utility of immersive interfaces.

This integration is probably best achieved by starting with a "top down/bottom up" design approach on a number of key applications. For example, the entertainment industry could use an immersive set design and preview system, while the Defense Department would benefit from a simulation-based design and modeling system that fully utilizes a human's ability to think, design, and manipulate 3D space.

PETER BONANNI

The U.S. armed forces have created the most advanced training systems in the world. Some segments of the armed forces, however, are facing true training shortfalls for the first time in decades. These training deficiencies are being caused by worldwide deployments. U.S. Air Force active duty and reserve squadrons, for example, have experienced a reduction in training sorties of up to 25 percent. This reduction is a direct result of deployments in support of contingency operations over Iraq and Bosnia. Pilots are most proficient and able to fight when they are first deployed to these areas. As the deployment wears on, with little or no training opportunities, pilot proficiency slips. The same problem is occurring in other combat arms as the trend to use U.S. forces in peacekeeping roles accelerates. Since conducting realistic training is impossible on most of these missions, simulators provide the only realistic training alternative. Unfortunately, most of the simulators in use today are very expensive, are limited to single-crew training, and are not deployable.

Emerging commercial simulation technology, however, may provide a near-term solution to this military training problem. Some fighter pilot skills, for example, cannot be practiced in simulation, regardless of the fidelity. The most important (and perishable) skills, however, can be honed by very-low-cost simulators. The computer game *Falcon 4.0* is an example of a commercial product that is shattering the fidelity threshold and providing a model for very-low-cost simulation. There are several key components to *Falcon 4.0* that allow this type of breakthrough. *Falcon 4.0* features "SIMNET-like" networking protocols that create a large man-in-the-loop environment. These features of *Falcon 4.0* provide the basic building blocks for producing a simulator that will be low in cost and deployable and that will provide pilots with team training opportunities. In the near term this capability will be enhanced with the development of commercial head-mounted displays and voice recognition systems.

DEFENSE MODELING AND SIMULATION OFFICE
DOD Modeling and Simulation Overview and Opportunities for Collaboration Between the Defense and Entertainment Industries

The U.S. Department of Defense (DOD) is building a robust modeling and simulation (M&S) capability to evaluate weapons system requirements and courses of actions; to reduce the time line, risk, and cost of complex weapons system development; to conduct training; and for realistic mission rehearsal. Part One of this paper provides a description of the current and envisioned application of M&S in the training, analysis, and acquisition support functional areas. It also summarizes the plan that is in place to help achieve DOD's M&S vision. Part Two is a list of technology areas that DOD believes have a potential for collaborative development with the entertainment industry.

PART ONE: DOD MODELING AND SIMULATION OVERVIEW

Vision and Application

The foundation for the above set of DOD M&S capabilities will be the development of a common technical framework to maximize interoperability among simulations and the reuse of simulation components. The cornerstone of the common technical framework (CTF), the High-level Architecture (HLA), has just been adopted as DOD-wide policy. Together with the other elements of the CTF, data standards, and a common understanding (or conceptual model) of the real world, the HLA will enable DOD to use and combine simulations in as-yet unimagined ways. Establishment of a commercial standard will follow as applications spread to training for natural disaster response, weather and crop forecasting, and a host of other business and social problems.

Common services and tools also will be provided to simulation developers to further reduce the cost and time required to build high-fidelity representations of real-world systems and processes. Realistic simulations, interacting with actual war-fighting systems, will enable combatants to rehearse missions and "train as we fight." Virtual prototypes developed in a collaborative design environment using the new integrated product and process development concept will be evaluated and perfected with the help of real war fighters before physical realizations are ever constructed. DOD will enforce recently approved policies and procedures for the verification, validation, and accreditation of models and simulations to ensure accuracy, thereby enhancing the credibility of simulation results.

The advanced M&S capability envisioned by DOD will be a rapidly configured mix of computer simulations, actual war-fighting systems, and weapons systems simulators geographically distributed and networked and involving tens of thousands of entities to support training, analysis, and acquisition. Not only is there a desire to quickly scale the size and mix of simulations, but DOD also is pursuing the capability whereby both groups and individuals can interact equally well with a synthetic environment. The major challenge in providing scalability, as well as the group and individual experience, is achieving consistency and coherence of both time and space.

Other areas of ongoing research in DOD that show promising results are the accurate representation of human behavior, systems, and the natural environment (air, space, land, sea, weather, and battle effects). DOD's efforts are focused on just-in-time generation of integrated and consistent environmental data to support realistic mission rehearsals anywhere in the world, including inaccessible or operationally dangerous locations. Investments in the rapid extraction of land and water surfaces, features existing on those surfaces, and features derived from ocean, air, and space grided fields have begun to yield results. The goal is to develop a capability to generate feature-integrated surfaces that require minimal editing and model-based software for feature extraction. Achieving this will, for example, ensure that weather fronts that bring rain or snow change the characteristics of the ground so that transit rate is affected and the associated wind patterns move trees, create waves, and alter dispersal patterns of smoke and dust. The resulting realism will add significantly to training, analysis, and acquisition. These effects, when coupled with dial-up capability to create custom correlated conditions, can provide year-round training.

Training

Warriors of every rank will use M&S to challenge their skills at the tactical, operational, or strategic level through the use of realistic synthetic environments for a full range of missions, to include peacekeeping and providing humanitarian aide. Huge exercises, combining forces from all services in carefully planned combined operations, will engage in realistic training without risking injury, environmental damage, or costly equipment damage. Simulation will enable leaders to train at scales not possible in any arena short of full-scale combat operations, using weapons that would be unsafe on conventional live ranges. Simulation will be used to evaluate the readiness of our armed forces as well.

The active duty and reserve components of all forces will be able to operate together in synthetic environments without costly and time-consuming travel to live training grounds. In computer-based training, both the friendly and opposition forces, or computer-generated forces (CGFs), are highly aggregated into large command echelons and carry out the orders resulting from staff planning and decision making. CGFs fall into two categories: (1) semiautomated forces (SAFs), which require some direct human involvement to make tactical decisions and control the activities of the aggregated force, and (2) automated forces, which are associated with autonomous agent (AA) technology. AAs are in early development phases and will find extensive applications in M&S as the technology matures.

There is now a diverse and active interest throughout the DOD M&S community, academia, and the software industry in the development of CGFs and AAs. The Defense Advanced Research Projects Agency is sponsoring the development of Modular Semi-Automated Forces for the Synthetic Theater of War program, which includes both intelligent forces and command forces. This effort also involves development of the command and control simulation interface language. It is designed for communications between and among simulated command entities, small units, and virtual platforms. The services, more specifically the Army's Close Combat Tactical Trainer program, is now developing opposing forces and blue forces to be completed in 1997. The British Ministry of Defence also is developing similar capabilities using command agent technology in a program called Command Agent Support for Unit Movement Facility. Academic and industrial interest in this technology has led to the First International Conference on Autonomous Agents, which will take place in Marina del Rey, California, on February 5-8, 1997.

Analysis

M&S will provide DOD with a powerful set of tools to systematically analyze alternative force structures. Analysts and planners will design virtual joint forces, fight an imaginary foe, reconfigure the mix of forces, and fight the battle numerous times in order to learn how best to shape future task forces. Not only will simulation shape future force structure, but it will also be used to evaluate and optimize the course of action in response to events that occur worldwide.

M&S representations will enable planners to design the most effective logistics pipelines to supply the warriors of the future, whether they are facing conventional combat missions or operations other than war.

Acquisition

Operating in the same virtual environments, virtual prototypes will enable acquisition executives to determine the right mix of system capability and affordability prior to entering production. Fighting synthetic battles repeatedly while inserting new systems or different components will help determine the right investment and modernization strategy for our future armed forces. Models and simulations will reduce the time, resources, and risks of the acquisition process and will increase the quality of the systems produced.

M&S will allow testers to create realistic test and evaluation procedures without the expense and danger of live exercises. "Dry runs" of live operational tests will minimize the risks to people, machines, and testing ranges.

M&S will enhance information sharing among designers, manufacturers, logisticians, testers, and end users, shortening the system development cycle and improving the Integrated Product Team development processes.

The DOD M&S Master Plan

The DOD M&S Master Plan (MSMP) is a corporate plan to achieve DOD's vision. Its first objective is the establishment of a common technical framework, anchored by the HLA. The HLA has been defined and adopted as the standard simulation architecture for all DOD simulations. Development continues on the other elements of the CTF, and DOD's investment strategy for M&S is focused on achieving the vision.

The second objective of the MSMP is to provide timely and authoritative representations of the natural environment. To this end, Executive Agents (EAs) have been established to coordinate development in their respective areas of oceans, aerospace, and terrain. EAs have also begun to explore potential commercial marketplaces for their databases.

The remaining objectives address representation of systems, human behavior, and establishing a robust infrastructure to meet the needs of simulation developers and end users. The infrastructure will include resource repositories—virtual libraries—and a help desk for users.

The final objective of the plan is to share the benefits of M&S. DOD must educate potential users about the benefits of employing M&S. To that end, an extensive study is under way to quantify objective data on the cost-effectiveness and efficiency of M&S in training, analysis, and acquisition applications throughout DOD. Extensive anecdotal data exist, but no concerted effort to demonstrate the return on investment has been done.

PART TWO:
M&S TECHNOLOGY AREAS

Although the vision for M&S described previously is focused on meeting the needs of the military, implementing the vision requires the development and exploitation of technologies that also have application to the entertainment industry. The following partial list of technologies was identified by members of the DOD M&S community as areas where cooperative development with the entertainment industry will have the greatest benefit to both communities.

Virtual Presence

Virtual presence is the subjective sense of being physically present in one environment when actually present in another environment (Sheridan, 1992). Researchers in virtual reality (VR) have hypothesized the importance of inducing a feeling of presence in individuals experiencing virtual environments if they are to perform their intended tasks effectively. Creating this sense of presence is not well understood at this time, but among its potential benefits may be (1) providing the specific cues required for task performance, (2) motivation to perform to the best of one's abilities, and (3) providing an overall experience similar enough to the real world that it effectively allows suspension of disbelief while at the same time the synthetic environment elicits the conditioned or desired response while in the real world.

Visual Stimulus

This is the primary means to foster presence in most of today's simulators. However, because of an insufficient consideration of the impact of granularity, texture, and style in graphics rendering, the inherent capability of the available hardware is not utilized to the greatest effect. One potential area of collaboration could be to investigate the concepts of visual stimulus requirements and the various design approaches to improve graphics-rendering devices to satisfy these requirements.

Hearing and 3D Sound

DOD has initiated numerous efforts to improve the production of 3D sound techniques, but it has not yet been effectively used in military simulations. Providing more realistic sound to a synthetic environment can have two potential benefits for training: (1) providing more realistic sound cues and (2) providing a more realistic aural environment that enhances realism.

Olfactory Stimulus

Smell can contribute to task performance in certain situations and can contribute to the full sense of presence in the synthetic environment. There are certain distinctive smells that serve as cues for task initiation. A smoldering electrical fire can be used to trigger certain concerns by individuals participating in a training simulator. In addition, smells such as hydraulic fluid can enhance the synthetic environment to the extent of creating a sense of enhanced danger.

Vibrotactile and Electrotactile Displays

Another sense that can be involved to create an enhanced synthetic environment is touch and feel. Current simulator design has concentrated on moving the entire training platform while often ignoring the importance of surface temperature and vibration in creating a realistic environment.

Coherent Stimuli

One area that has not received much research is the required coherent application of the above-listed stimulations to create an enhanced synthetic environment. Although each stimulation may be valid in isolation, the real challenge is the correct level and intensity of combined stimulations.

Virtual Environment Representation

This area includes technologies that emphasize the representation of individuals and the interactions among virtual and live participants in an individual or group experience.

Representation of Human Figures

While methods are evolving for creating computer-generated representations of human figures that are anthropometrically valid, in general these methods are computationally complex while at the same time stylishly rigid. Approaches for automated modeling of human figures that result in more natural representations that are more computationally efficient is a topic of great interest in a number of disciplines, including medicine. The need is to determine the minimum essential information required to provide a representation of human actions that are sufficiently realistic for both communities. Animation of human figures, including speech, running, and facial expressions, still requires significant development.

Human Body Tracking

Research has begun on methods for tracking and capturing the motion of humans that support real-time interaction with both virtual and constructive simulations.

Virtual Backgrounds

Creation of a full virtual environment requires generating the natural and/or cultural features of the background in which the interaction takes place. Specific areas of research include automating the production of background environments and efficient representations in scalable databases.

Group/Team Interactions

Most of the research in virtual presence has been single person oriented (e.g., head-mounted displays and tracking systems, hand and foot controls). DOD has a direct interest and experience in developing the group or team training experience, which is also of interest to the entertainment industry. DOD would like to enhance its capability for an entire group to interact with a virtual environment and each other without the need for unique individual hardware devices.

SUMMARY

The DOD vision is to apply M&S to the full range of military applications, including training, analysis, and acquisition. The vision can only be met if the technology defined above is readily available, of low cost, and operationally valid. It is the desire of DOD to explore technologies with the entertainment industry that are relevant to modeling and simulation. These technologies may include animation, graphical imaging, data communication and storage, architectures, and human immersion. DOD believes research in collaboration with the entertainment industry will provide mutual benefit to both communities.

REFERENCE

Sheridan, T.B. 1992. *Telerobotics, Automation, and Human Supervisory Control.* MIT Press, Cambridge, Mass.

JAMES FORAN

Nintendo 64, the first truly interactive three-dimensional video game machine, provides a level of experiences that has not generally been available outside the traditional simulation and training community. It does so at a price point that allows virtually every household to own one. The implications of the technology embedded in the machine for all types of training and simulation are tremendous. Not only does it provide a low-cost ubiquitous platform, but it also portends a future where even more powerful and realistic machines will be pervasive.

Silicon Graphics, relying on 15 years of experience, was able to utilize state-of-the-art semiconductor technology to achieve a low-cost, high-performance, high-volume product for Nintendo. The chips utilized were among the first logic chips to be produced using 0.35-micron technology. This represents a fundamental change in the way technology is driven. In the past, advanced technologies were first used to produce low-volume, high-cost systems principally for military use. These seed applications provided the opportunity to make the technology viable economically. Over time the technology moved down in product price point until eventually it appeared in consumer devices.

All of this has now changed. Today, with fabrication facilities costing over $1 billion, large-scale markets must exist to justify the expense of construction. Although DRAM [dynamic random access memory] has long been the primary justification for new fabrications, the cyclical nature of demand requires that other applications need to exist to balance capacity utilization. Video games are the largest market for consumption of advanced semiconductor technologies; their public acceptance is orders of magnitude higher than that of traditional computer products. In its first six months, 2.7 million units are expected to be sold, increasing to a total of 5 million within the first nine months.

In the video game market it is possible to get an advanced product like this out at a price point that is acceptable to the consumer only because it is possible to subsidize hardware with software. The hardware is brought to market with a very low margin throughout the chain from manufacture to distribution. Much like a CD player, the box has no intrinsic value to the consumer; it is simply a necessary expense in order to enjoy the game. Over the product life it is typical for each platform to average 10 games. This provides the return on investment to the manufacturer as well as a living for the content providers.

This is a great development for kids who want to play games, but what implications does it have for other markets? It is instructive to look at some of the similarities to the requirements that are traditionally associated with the military market. Typical military programs have stressed advanced technology. After all, competitiveness is the cornerstone of any military development. The video game business is a war for the consumer pocketbook. Because of the requirement for competitive advantage, both applications are up-front and capital intensive. Long-term product stability also is important for both markets. In this respect, video games are unusual for a consumer product. Each hardware unit in the field must play every game cartridge the same as every other machine. Maintenance of the design for a 10-year period is accepted.

So we can see that there are many characteristics of video game hardware that match up with typical military requirements. How could this type of hardware be put to use? In the field of training and simulation the military has long led the way. With increasing sophistication of weaponry and the political sensitivities associated with the type of actions encountered in today's world, military preparedness is more necessary than ever. Simulation also provides the cost-effectiveness required by today's budget realities. Nevertheless, practical training equipment, although decreasing in price, has not yet become ubiquitous. This type of video game platform now makes that possible.

The question before this group is, How can the military take advantage of this commercially developed technology? One immediate answer is that training cartridges can be developed for the actual home game platform. This requires the setting up of some sort of development rights with the game platform manufacturer. This is actually a very practical method for training applications where the home game hardware is sufficient to achieve the training objective. In the case where input devices must be similar to actual operational hardware or where systems must be embedded into operational equipment, one must go beyond the box available at the toy store.

Some of these requirements can be met by physically reconfiguring the hardware and developing the appropriate accessories. In other cases, where requirements may exceed the capabilities of the home game box, more powerful systems can be built utilizing the same components.

Generally, a semiconductor process yields a speed range of parts that can vary in horsepower by 50 percent or more. In the case of a product like Nintendo64, because of the requirement for high volume and low cost, all devices must work in the target system. This means that through speed grading much more powerful components can be obtained. By using these components and more robust system configurations it is possible to satisfy more demanding requirements. Since the semiconductor process used to manufacture these state-of-the-art devices is itself quite new, it is a natural that as the process matures a shift of yield to higher-speed parts will result. This is a quite common phenomenon in the DRAM business.

So what is the issue that prevents this type of technology from being utilized by the military? The military can accomplish tremendous projects during times of war or national emergency, but during peace time the design and procurement cycle moves at a snail's pace. I recently talked to a customer—a military system integrator—about designing a graphic function for use in a new vehicle. He was concerned that he might prototype with something that would not be cost-effective in implementation. I asked him: "Well, how far out is production?" The answer was that production would start in six to seven years. I told him there wasn't anything on Earth with regard to electronics that would not be cost-effective in six years if it exists today.

How can the military deal with this situation?

1. *It can think long term.* We have to have a vision of what kind of technology we will want to be using 5 to 10 years from now. We have to be practical. There are far too many "futurists" on the speaking circuit whose ideas are either too far out in time or lack any understanding of the infrastructure required to provide a whole solution. Nevertheless, a long-term vision is a necessity.

2. *It can focus.* Focus on specific objectives of large scale. This provides volume, which is necessary to entice companies to commit resources. However, don't get trapped into trying to define a universal device that meets all needs for everyone. The desired product must be simple to describe and easily understood by everyone involved.

3. *It can make commitments.* No company with shareholders is going to make a major investment of its resources for something that might happen. The military must realize that despite everyone's good intentions some of these commitments will result in failure. Monitoring of progress and conditions is required, but this needs to be at arm's length.

4. *It can cut the red tape.* Companies need to be able to do what is business as usual to them. Requirements need to be clearly communicated and feasibility agreed. After that, get out of the way and focus on results rather than procedures.

5. *It can put the infrastructure in place.* No solution will succeed unless the proper infrastructure is put in place on the customer side. There needs to be honesty in the assessment of how fast infrastructure changes can be accomplished, and these must mesh with the time scale of the project.

In closing, there is no limit to where technology can go. The limits lie in our ability to apply technology. The reason that entertainment markets are able to apply technology successfully is that major commitments are made for specific focused objectives of large magnitude. This provides lucrative opportunities for technology companies to provide new solutions.

JOHN GEDDES
Collaborative Challenges:
Understanding Cultural Differences

There is a broad cultural chasm between the U.S. Department of Defense (DOD) and the entertainment/computer industry. This chasm can present a serious obstacle to successful collaboration. Processes and attitudes will have to be created or modified if collaboration is going to succeed between the two groups.

In the context of modeling and simulation, DOD can be characterized by varied and often competing interests, funding that is renewed annually, and extremely hierarchical and time-consuming approval and review processes.

- *Varied and competing interests.* Three domains of simulation competing for funding (ACR, RDA, TEMO); uncertainty and competition between DARPA, RDECs, STRICOM, and major commands like SSDC for primacy in development and program management of new simulation activities.
- *Funding uncertainty.* Annual budget processes, effects of changing military and civilian leadership on priorities; for example, Army Modeling and Simulation organizations (MISMA, AMSO, DUSA/OR, M&S GOWG) and National Rotorcraft Technology Center funding profile.
- *Need for coordination across commands and agencies* to get approval and requirements for periodic reviews at multiple levels; examples in ACTD processes, Soldier System development. Long duration of projects—one year to get consensus, two years until funding; examples in Louisiana Maneuvers, Battlefield Visualization.

In comparison, the commercial entertainment/computer industry can be characterized by short project horizons, more stable funding, relatively flat heirarchies for approval, and more informal and spontaneous review processes.

- Product horizons are one to three years from concept to product on the shelf; an example is the Nintendo 64.
- Once a company approves a project for development and production, funding is programmed and maintained generally for the duration of the effort and is not subject to the whims of elected representatives.
- Flatter hierarchies and more informal reviews, resulting from total quality management or reengineering and closer scrutiny of value-added functions; less internal regulation.

Recommendations for successful collaboration:

1. Create an advisory board with power to sponsor and recommend collaborative and cooperative efforts. Publish annual report with positive results and with opportunities that were neglected. Include lessons learned about positive and negative collaborative results.

2. Exchange liaisons. Create positions that are geographically proximate for providing effective coordination and for seeking opportunities—a few that work for advisory board, more that work for specific participants, both DOD and non-DOD.

3. Allow decisions at the lowest levels. Minimize hierarchical reviews. Nonproductive time for most participants. Use advisory board liaisons.

4. Understand and use existing cooperative mechanisms—cooperative research and development agreements, cooperative agreements, and other transactions. Involve a congressional staff in advisory panel to help shape future mechanisms.

MARTIN GUNDERSEN

Advanced modeling and simulations for games, entertainment, manufacturing, education, the U.S. Department of Defense, finance, and other applications will grow with the development of integrated media systems incorporating software and hardware development at many levels. Integrated media systems will powerfully impact all fields of inquiry and technology. Integrated media systems of the future will seamlessly combine digital video, digital audio, computer animation, text, and graphics into common displays in such a way as to allow for mixed media creation, dissemination, and interactive access in real time. Prodigious national and international resources are currently being marshaled for integrated media technologies' research, development, infrastructure installation, product creation and commercialization, public performance, and training. According to a recent projection, multimedia and creative technologies will represent a new total market of $40 billion by the year 2000 and $65 billion by the year 2010.

The beckoning opportunity is to accelerate progress in this new discipline by revolutionizing our access to information sources, easing the effort required to author original works, and transforming our capacity to augment and enhance the productivity of human creative endeavors. The corresponding challenge is to first recognize the impact of these dramatic changes on the very nature of our teaching tools, manufacturing methods, defense systems, health care systems, and entertainment/art forms and to then exert sufficient positive leadership to assure maximum benefit. At the University of Southern California we are pursuing a large-scale program that is relevant to the goals of utilizing entertainment-oriented technology. We have established a Center for Integrated Media Systems, which is directed by Max Nikias, for research, development, and teaching in advanced systems for multimedia applications, including entertainment. This research has recently received one of four Engineering Research Center awards this year from the National Science Foundation, the proposal ranking first out of 117 proposals.

There are three major areas of importance with opportunities for research and development: interfaces, communications, and databases. These are discussed below. The next generation of integrated media systems in the augmented reality, interactive multimedia, heterogeneous computing, distributed database, wireless communication, and high-speed network environments will impact every facet of our lives. Access to a wealth of diverse and distributed information resources will be possible from within an individualized "information framework." Interactive media will enable new paradigms for education, training, manufacturing, and entertainment that provide worlds—real, augmented, and fantasy—for people to experience, learn through, and interact with. Design-based industries will develop products through virtual design systems that integrate software applications and manage both the design process and the design data, as well as incorporate input from intended consumers, designers, production engineers, quality assurance and quality control specialists, cost analysts, and manufacturing engineers.

CREATOR-COMPUTER-CONSUMER INTERFACES

Computer interfaces are unidirectional and inefficient. A significant bottleneck has emerged at the creator-computer and computer-consumer interfaces owing to an increasing mismatch between computational and display power, on the one hand, and human-computer input/output (I/O) on the other. Simply put, highly visually and aurally oriented human beings are constrained to interact with an assistant that cannot see, hear, or speak. The human-computer interface has evolved over four decades from plug boards, lights, punch cards, and text printers to postscript laser printers, mouse-based window systems, and primitive head-mounted displays. The trend has clearly been toward interaction modes that are more intuitive, enabling people to communicate more effectively to and through computer systems. Today, enhanced video and audio capabilities fuel the explosive success of both multimedia-equipped studio-grade workstations (the creator-computer interface) and personal computers (the computer-consumer interface), as particularly evidenced by the trend toward truly interactive media applications.

Technological advances in the area of human-computer interfaces are necessary to achieve a new level of even richer and more perceptive interfaces that are characterized by the *immersion* of the user/participant in highly communicative multisensory interactions. These advances must span both visual and aural interface technologies. Input to the computer can be enhanced by means of *smart cameras* for environmental awareness and expression recognition and with *robust speech recognition* for extended natural language interactions. *Immersivision* methods for panoramic scene reprojection and novel approaches to *three-dimensional (3D) displays* enrich the presentation of graphic output. The computer's sense of the environment is enhanced through smart-camera-based *tracking technology*, which in turn is pivotal for both *augmented reality applications* and the synthesis of an accurate 3D aural environment through *immersive sound reproduction*. Furthermore, the coupling of these technologies with advances in wireless networks and distributed databases will allow the integration of mobile workstations (personal data assistants) with tracked head-up displays for application in augmented office, classroom, factory, and cockpit environments.

MEDIA INTERCONNECTION AND DELIVERY FABRICS

Real-time distribution and storage of multimedia information is expensive. Even with compression, which can only be employed in certain applications, digital video and audio can consume large portions of database storage and network bandwidth. Access to even currently available network bandwidth is limited by workstation I/O design bottlenecks. A need therefore exists for both high-bandwidth interconnections and interfaces and real-time artifact-free compression and decompression algorithms.

Over the past decade, user demands on networks and databases have escalated from the bandwidth and storage requirements characteristic of text to those characteristic of both images and real-time production-quality video and audio. As integrated media systems evolve to incorporate the advanced interfaces described above, they will impose even greater demands on high-speed wired and wireless communications networks. These enhanced visual and aural interfaces, as well as real-time digital video servers, integrated media databases, and distributed processing systems will require the effective and efficient image and data compression methods, multi-gigabit-per-second (Gbps) fiber-optic networks, and high-bandwidth wireless networks developed in this thrust. Two cases illustrate how the need for such delivery fabrics arises depending on the number of connected users. In today's manufacturing environments with hundreds of untethered workers, or in video-on-demand networks with thousands of consumers, each person requires on the order of 20 Mbps of bandwidth over wireless or wired networks to receive compressed video and graphics. On the other hand, in today's video production environment with dozens of users, each requires about 270 Mbps for D1 digital video. A shared network is an efficient means for distributing data in both of these cases. One challenge for such a system with multi-Gbps (2 to 50 Gbps) aggregate throughput is to seamlessly support multiple data types such as D1, MPEG, text, and graphics. In addition, the interconnection and delivery fabric must be capable of satisfying future standards, such as video quality that is significantly superior to that of D1 or high-definition television. The research challenge in this area is focused on the development of technologies for shared integrated media networks.

DISTRIBUTED MULTIMEDIA INFORMATION MANAGEMENT

An effective methodology for managing large integrated media databases does not exist. Integrated media databases of the future will contain terabytes of information. Information relevant to a given need will likely reside in a collection of interconnected heterogeneous and distributed knowledge bases. Techniques for locally organizing, browsing, discovering, and querying such integrated media repositories are needed. Furthermore, many applications demand seamless synchronous access to multiple audio and video threads from distributed digital databases, a capability that does not currently exist.

Advanced human-computer interfaces and enhanced wired and wireless media interconnection and delivery networks cannot function effectively without access to dramatically scaled-up databases that can seamlessly manage multiple media types. Hence, the central integrated media-systems-related issue that must be addressed during the next decade is the storage, indexing, structuring, manipulating, and "discovery" of integrated multimedia information units (MIUs) that include structured data values (strings and numbers), text, images, audio, and video. The key research focus in this area centers on managing multimedia information units in the context of a highly distributed and interconnected network of information collections and repositories. Current data and knowledge management technology that addresses collections of formatted data and text is inadequate to meet the needs of video and audio information, as well as the mixture of modalities in MIUs. Furthermore, the highly distributed and interconnected nature of the emerging information superhighway accentuates the need for techniques that enable multimedia information sharing. The research challenge in this area involves the development of mechanisms that address four critical aspects of distributed multimedia information management: (1) multimedia information content representation and extraction; (2) multimedia database networking: discovery, filtering, query, sharing; (3) storage of and access to continuous media data types; and (4) visual presentation of information across cultures.

We are developing collaborations with other efforts, including related research activity at Howard University and the University of South Carolina. The South Carolina program has initiated development of a "virtual testbed," which is a top-down, mission-oriented approach emphasizing simulation of complete electrical systems on U.S. Navy ships using advanced visualization techniques. This program is under the direction of Professor Roger Dougal.

At the University of Southern California we have developed an industrial partnership with over 50 companies that are literally a cross-section of industry working to develop and apply the new technology. In entertainment we have formed a panel of entertainment professionals who will foster collaboration with the Hollywood industry that will be strongly impacted by multimedia simulations and modeling. The professionals are a cross-section of the industry, including actors, directors, film editors, audio engineers, computer network experts, writers, and others, including investors. Over the next few years we will be working to provide an academic venue for this technology to be researched, viewed, and understood, with emphasis on entertainment applications. The panel on entertainment applications will be meeting with industrial partners of the center at USC in a review that will occur in November of 1996.

WILL HARVEY
The Future of Internet Games

Latency is a major barrier to fast-action Internet games. Game developers can either hope the problem goes away or adopt new game architectures that work around it. There is compelling evidence that the problem will never go away and that the hardware will never improve to the point that developers can afford to treat the Internet like a local area network (LAN). Sandcastle offers an alternative, a software solution that enables fast-action Internet games.

High latency is incompatible with the client/server and lockstep designs that current LAN games use. A response time of 33 milliseconds (ms) has been the industry standard for over 20 years, and even with premium on-line services, Internet performance is nowhere near that level. In fact, it *cannot* reach that level. In fiber, light takes 54 ms to travel roundtrip between New York and San Francisco. Networking experts agree that the Internet's latency will plateau between 100 and 130 ms cross country (Figure D.1).

Fast-action client/server and lockstep games are no fun at this speed. A player trying to dodge a bullet will feel either frustrated, because the response time is too slow for him to dodge, or cheated, because the program displays his character such that it appears he has dodged when he has not. Punches a player could land will miss; opponents a player could tackle will evade. Without responsiveness, fast-action games are not fun.

THE SHIFT TO DISTRIBUTED PROCESSING

The solution is to move to a distributed architecture. In a distributed game, each player controls a character on his local machine, so it responds to his actions instantly, with no latency. The new challenge is then to synchronize the game state on all the machines and to coordinate interactions among objects that different players control.

In Figure D.2 the big circle is a server or multicast router in a building. The small circles are machines in people's homes. X, Y, and Z represent objects controlled by users from their own homes. Proxies not shown in these figures display the objects on every machine. The X, Y, and Z letters represent the point of control of each object.

In the lockstep architecture, each machine broadcasts its user input to the other machines and advances one simulation cycle when it has received a complete set of user input from all participating machines. Since advancing a cycle requires complete exchange of user input, the responsiveness is limited by the speed of the worst communications la-

APPENDIX D

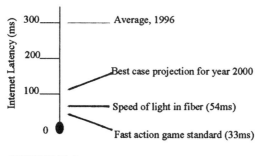

FIGURE D.1

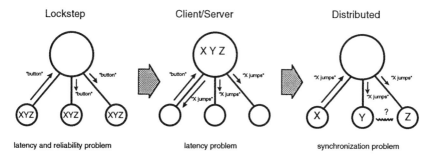

FIGURE D.2

tency of any machine. In the client/server architecture, each machine independently sends its user input or action request to the server in order to perform an action in the simulation. Controlling an object from a client machine still entails a roundtrip delay, but the responsiveness of any individual client machine is not affected by the communications speed of the other machines. In a distributed architecture, machines control objects locally and broadcast the results of actions to other machines, which receive the information with some time delay. Each machine has immediate responsiveness controlling its own objects but must synchronize interactions between its own objects and objects controlled by remote machines.

In both the lockstep and client/server architectures, responsiveness is limited by the roundtrip communication latency to the server, which will always be too long for fast-action games. Controlling objects locally and synchronizing interactions between them is the only solution. The shift from a central architecture to a distributed architecture transforms the latency problem into a synchronization problem.

SOLVING THE SYNCHRONIZATION PROBLEM

Methods of solving the latency and synchronization problems fall into three categories, represented in Table D.1.

At the lowest level, the first approach attempts to improve the speed of the network to reduce the latency problem by brute force, instead of adopting a distributed architecture. This approach will always have slow reactions because of the speed of light and network overhead, so it will be limited to domains like Quake, where players don't have a true opportunity to dodge bullets.

The second approach follows a software technology called distributed interactive simulation developed for military simulations. This approach accommodates the delay in which information is received from other participants by "dead-reckoning" or predicting the actions of the other participants to bring all objects displayed on a machine into the same time frame. Because predicting only works for predictable and continuously moving objects, such as planes and tanks, it does not apply to domains of rich human interaction like playing Nintendo's Mario 64 or playing catch with a ball over the Internet.

The third approach, synchronization, leverages off of the other two technologies, but more importantly it picks up where the other technologies reach their fundamental limitations. Information from remote machines will *always* be received with a time delay, and many actions *cannot* be predicted. Thus, remote objects *must* be shown in time delay. If a user has no interactions with remote objects, he cannot tell that he is seeing those objects "in the past"; but if he does interact with them, those inter-

TABLE D.1

Description	Technology	Latency	Limitations	Enables
Improve network—to reduce discrepancy between time frames	Premium service	150 ms	Slow reactions	Guns
Dead-reckon—to pretend objects are in the same time frame by predicting their positions	DIS and derivatives	0 ms perceived	Limited to predictable domains	Planes, tanks
Synchronize—interactions between objects in different time frames	Sandcastle	0 ms perceived		Picking up an object, catching a ball

actions must accommodate the time difference. Synchronization technologies are a set of software networking components that enable interactions between objects in different time frames.

SANDCASTLE'S DIRECTION

Sandcastle is developing synchronization technologies that give users the impression that the network has zero latency, or immediate responsiveness. Specifically, the technologies address the problems of interacting with shared objects, like throwing a football between users, and interacting directly with objects controlled by remote machines, as in a fighting game or a race.

Our view is that the latency problems of central processing are fundamental. Over time, the demands for high responsiveness will drive an inevitable shift in programming paradigms from central processing to distributed processing. As this shift occurs, the technologies and tools that address the critical problems of real-time distributed applications will become increasingly important. We believe that these contributions are the beginnings of a foundation not just for games and chat environments but for all of twenty-first century interactive entertainment.

ROBERT JACOBS

As the Computer Science and Telecommunications Board (CSTB) of the National Research Council assesses research priorities for defense and entertainment simulation, it must be mindful of the significant differences in objectives, risk and reward environment, and business traditions and customs, especially with respect to proprietary intellectual property, that characterize these two simulation industries.

Defense simulation programs focus on the solution of problems, the production of operational skills through training, the support of combat development test and evaluation, or the resolution of complex engineering optimization questions as a part of design and development. How well a defense simulation achieves its mission is usually determined by how its designers tailor the technology to address the problem of interest. Entertainment simulations, on the other hand, are a medium for the delivery of recreational experiences; the measure of success is not a matter of problem solution or production of information or skill but rather is determined largely by how exciting and enjoyable the experience is for the paying customer. The "fun quotient" of an entertainment simulation is predominantly a matter of art rather than technology; the technical side of the system must be capable of presenting the "story," but the perceived value of the experience hinges largely on the quality of the creative element.

Defense simulations are developed to a specification that defines the nature of the virtual world and the expectations of the customer for behaviors to be executed within it. As long as the product meets the specification, the development is deemed a success. Developers are compensated on the basis of their development cost plus a modest margin whose magnitude is negotiated in accordance with guidelines reflecting whether the customer or the developer takes on the development risk. There is no end-user specification to be met for an entertainment simulation. The developer must identify a market need, formulate a creative concept that addresses that need, and then back his intuition by investing his own money to field the concept. Maybe the marketplace will accept the concept; maybe it will reject it. For the most part, the market has been disappointed to date. If the simulation sells, it is priced in accordance with what the market will bear in view of competition, useful economic life, perceived value, return on investment, and so forth. What the market will bear may or may not be enough to recover development costs and to realize an attractive margin.

When the government contracts for research and development, both the client and the contractor generally acknowledge that the product of the effort belongs at least in part to the client. In the best case (from the

APPENDIX D

contractor's point of view), the developer may share in the right to future exploitation of what is produced; however, the government belongs to all of us, and the government's equity in the ideas and technology is part of the public domain. In the entertainment world, proprietary intellectual property is the principal stock in trade, and ownership of the right to future exploitation is the primary asset resulting from the investment in a project. The customer buys the right to exhibit the product but never the right to the underlying proprietary intellectual property. Technological and creative innovations are important contributors to the asset value of the enterprise. To the extent that they can be protected, they will not be willingly given away.

WHERE IS IT ALL GOING?

These contrasts, particularly the last point, create an interesting challenge for the CSTB in its quest to encourage open collaboration between defense and entertainment simulation developers. There is such a difference in the operational norms between these two industry segments that the resulting cultural barrier has been successfully breeched in only very few instances. One might expect that there also is a divergence of views as to how the industry and its technology will evolve in the coming decade.

On the defense side, the next few years will see continuing efforts to develop and disseminate technologies for more effective application of simulations to military and civil problems. These will include:

• Increased emphasis on large-scale simulations of military activity at the joint and coalition levels.
• Increased dependence on simulation technology to offset cuts in OpTempo, to conduct distributed planning and rehearsal, and to provide visualization for distributed command and control.
• Increased ubiquity of simulation, so that players will be able to join distributed virtual activities from any place and at any time.
• Increased capability for scalability—from combat theater to foxhole—with appropriate level of detail to support activity at either extreme.
• Improved ability to represent the behaviors of forces by computer-driven virtual entities to include complex concept formulation, planning, and reasoning activities in addition to simple drills.
• Increased availability of communication bandwidth to accommodate more simultaneous players, accommodate demand for more tightly coupled and reactive simulation processes, to realistically stress players, and to realistically simulate "fast" processes.
• Increased availability of tools for economical "rapid prototyping."

In the entertainment simulation world, return on investment is a key consideration. Research and development will focus on achieving value in the perception of the end customer. The need to impress end customers whose experience base is grounded in television and the real world will focus the competition at the highest levels of fidelity consistent with economic pricing. Pressure will continue to increase the performance and reduce the cost of leading-edge technologies so that each new generation of a product stimulates new demand and creates a competitive edge over its predecessor.

A conflict can be expected to develop between advocates of open standards and guardians of proprietary intellectual property. The substantial barrier to entry represented by development investment and the reduction of same that common standards promote will be cited by both groups as justification for promoting or avoiding the adoption of technologies common to competitive development teams. Ultimately, competition will refocus on the creative aspects of entertainment simulations, as developers realize that economy and speed in bringing an idea to market are greater factors in economic success than proprietary technology.

Entertainment developers suffer an approach-avoidance conflict over the accelerating pace of technological innovations, both because of the diminishing economic half-life of a development investment and the chaos in the competitive environment that the continuing avalanche of new capabilities will create. Even savvy buyers will become dizzy and indecisive as great products are eclipsed by pending spectacular ones.

ENABLING TECHNOLOGIES

What are some of the research priorities that will fuel the evolution suggested here? (1) Continuing geometric advancement in computing power, especially in the special-purpose hardware that creates imagery, with an accompanying dramatic reduction in price per performance. We can look ahead to the availability of photorealistic interactive systems at a price affordable by every household—e.g., the cost of a television set. (2) Dramatic improvements in the capability to display virtual environments to human senses: very-high-resolution visual displays; true spatial sound; and tactile displays that communicate surface qualities (friction), resilience, and thermal characteristics (heat capacity).

TRACI A. JONES

There has recently been an increased focus on simulating and modeling the individual soldier within the synthetic, or virtual, battlefield. The U.S. Department of Defense (DOD) has approved a Defense Technology Objective (DTO) for Individual Combatant Simulation (ICS). The ICS DTO is currently supported by an Army Science and Technology Objective (STO) for ICS. This is a joint STO between the Simulation, Training and Instrumentation Command (STRICOM) and the Army Research Laboratory, coordinated with the Natick Research and Engineering Directorate. The program intends to procure and demonstrate technologies for creating real-time simulations to immerse the individual soldier and allow for interaction in a synthetic environment. The cost-effectiveness of networked virtual reality devices will be determined using a multisite distributed laboratory consistent with DOD's High-level Architecture. The STRICOM Engineering Directorate is working closely with the Project Manager for Distributed Interactive Simulation (DIS) on the Dismounted Warrior Network project, which will take advantage of several technology-based efforts to provide an engineering proof of principle for immersing an individual into a synthetic environment.

The products that will evolve within DOD include the definition of a systems architecture to support the requirements for ICS as well as platforms and simulations that will support low-cost capabilities for mission rehearsal, materiel development, and training of individual soldiers and marines. There also is potential application of these technologies to training and rehearsal for the Federal Bureau of Investigation and the law enforcement industry.

The technological advances required and the technological challenges include low-cost solutions for:

• Visualization of human articulation in real-time networked environments,
• High-fidelity fully immersive systems,
• Interoperability between different fidelity simulators,
• Expansion of computer-generated forces for intelligent individual soldier interaction and decision making,
• Integration of high-resolution terrain databases with immersive simulations instrumentation of the individual for high-precision engagement data collection capability within buildings,
• Rapidly generated terrain databases to support mission planning and rehearsal while en route to a conflict, and
• Accurate simulation of weapons systems in real-time computer-generated environments.

Regarding complementary efforts in the entertainment or defense sectors that might be applicable to my own interests, I am aware of the motion-capture techniques used by the entertainment industry, primarily for game development and motion picture special effects. One such product is being used for the STRICOM Dismounted Soldier Simulation (DSS) system, under contract to Veda Inc. DSS uses a wireless optical tracking system developed by the Biomechanics Corporation for Acclaim Entertainment. The technology has been integrated into a real-time DIS environment. The untethered soldier, outfitted with a set of optical markers and wireless helmet-mounted display, moves about freely in a real-world motion-capture area, while position and orientation data are gathered and sent to a DIS network via tracking cameras and image-processing computers. Fully articulated human motion rotations and translations are sent out to the DIS network using entity state and data protocol data units. Issues such as network bandwidth limitation and system latency have been analyzed.

Other potential products being developed by STRICOM have application to the entertainment industry. The Omni-Directional Treadmill is an example of a locomotion simulator that allows an individual to walk and run in a virtual world. As the user moves on the treadmill, his view of the computer-generated world changes, immersing him into the virtual environment. The Army may use this technology, for example, to rehearse for a mission by walking through a hostile environment beforehand. It is anticipated that additional technologies developed by the entertainment industry can be leveraged to support DOD requirements for individual combatant simulation.

EUGENIA M. KOLASINSKI
Predicted Evolution of Virtual Reality

As this report indicates, virtual reality (VR) technology has many promising applications in both the simulation and entertainment arenas. VR technology is already being used for simulation, and, as the cost decreases, its many potential applications will likely lead to widespread use of VR, especially in the home for entertainment.

NECESSARY TECHNOLOGICAL ADVANCES AND PRIMARY RESEARCH CHALLENGES

Unfortunately, a phenomenon exists that may pose a threat to the ultimate usability of this new technology. That phenomenon is referred to as "simulator sickness" and it is a well-documented effect of simulator exposure (Reason and Brand, 1975; Kennedy and Frank, 1983; Kennedy et al., 1989; Casali, 1986). Simulator sickness is similar to motion sickness but can occur without actual physical motion. The cardinal signs resemble those of motion sickness: vomiting, nausea, pallor, and cold sweating. Other symptoms include drowsiness, confusion, difficulty concentrating, fullness of head, blurred vision, and eye strain. Along with the potential discomfort to the individual, there are several operational consequences of simulator sickness: decreased simulator use, compromised training, and ground and flight safety (Crowley, 1987). There are additional effects of simulator exposure: delayed flashbacks and aftereffects (a sudden onset of symptoms) (Baltzley et al., 1989); shifts in dark focus (the physiological resting position of accommodation) (Fowlkes et al., 1993); eye strain (Mon-Williams et al., 1993); and performance changes (Kennedy et al., 1993).

One potentially critical effect of simulator exposure is postural disequilibrium, referred to as *ataxia*. Baltzley et al. (1989) suggested that unsteadiness and ataxia are the greatest threats to safety because there have been reports of such posteffects lasting longer than 6 hours and, in some cases, longer than 12 hours. Clearly, occurrence of ataxia has the potential for disastrous consequences.

Recent research (Kolasinski, 1996; Knerr et al., 1993; Regan, 1993) has documented that simulator sickness can also occur in conjunction with VR exposure. The potential consequences of such sickness—particularly with widespread use of VR technology—raise important safety and legal issues for both manufacturers and users alike. Thus, simulator sickness (including effects such as ataxia) as it occurs with VR exposure must be understood if the technology is to make its predicted progress over the next decade. To meet this goal, the primary research challenges will be to thoroughly investigate the phenomenon.

Fortunately, simulator sickness in a virtual environment (VE)—or "cybersickness," as it is called—need not be regarded as an entirely new phenomenon. As already noted, simulator sickness is related to motion sickness, a phenomenon for which a body of literature exists (Reason and Brand, 1975). In addition, a body of literature exists for simulator sickness occurring in military flight simulators and, to a lesser degree, other simulators such as driving simulators (Crampton, 1990). Thus, VR researchers need not entirely reinvent the wheel but can and should draw on the existing literature, at least in the initial stages of investigation.

Much of the sickness literature that may be applicable to VEs is reviewed by Kolasinski (1995). In this report, three major categories of factors that may be related to simulator sickness as it occurs in a VE were identified: factors related to the individual using the system, factors related to the task performed in the VE, and factors related to the VR system itself. Although simulator sickness is not a new phenomenon, a VE may differ in several important respects from the typical simulator. For example, depending on how a VE is defined, such a system is likely to involve some form of direct sensory input, probably through a head-mounted display (HMD), at least. Such devices may pose unique concerns, and current research efforts (Mon-Williams et al., 1993) are examining the effects of HMD use on the visual system. Thus, although research into sickness occurring in VEs can draw on previous simulator sickness research, new research must be conducted specifically in VEs in order to address sickness issues unique to the VR setting. Very little research exists on sickness as it occurs in conjunction with VR exposure. Furthermore, with few exceptions (Regan and Price, 1994), the majority of VR studies currently reported in the literature were not designed to specifically investigate sickness. Instead, most studies investigated the use of VR systems, with sickness examined only as an aside.

Kolasinski (1996) represents one of the first experimental investigations of simulator sickness as it occurs in VEs. The primary focus was to investigate the prediction of sickness based on characteristics associated with an individual using a VR system, but the occurrence of ataxia following exposure also was investigated. This research established that sickness did, in fact, occur. In some cases it was severe—one participant vomited—and/or involved lingering or delayed effects. Ataxia, however, was not found.

This latter finding—that ataxia did not occur even though sickness did—supports findings presented by Kennedy et al. (1995), who found that, with repeated exposure to a simulator, sickness decreases over time but ataxia increases. Although their finding has implications for repeated use of VR technology, the finding of Kolasinski (1996) raises some specific issues of importance to the future application of VR technology. Ataxia is a well-documented effect of simulator exposure (Kellogg and Gillingham, 1986; Kennedy et al., 1993), and previous research has suggested that ataxia may also occur in conjunction with VR exposure. Rolland et al. (1995) found degradation in hand-eye coordination and errors in pointing accuracy following the wearing of a see-through HMD—results that demonstrate that negative aftereffects are indeed possible. There have also been anecdotal observations of individuals demonstrating significant ataxia following a 30-minute VR exposure (K.M. Stanney, personal communication, April 9, 1996). Finally, recent research (Kennedy et al., 1996) has concretely established the occurrence of ataxia following VR exposure.

The VE used in conjunction with the anecdotal observations referred to above was a maze, the traversal of which involved both forward and left/right-represented movements. On the other hand, the task employed in Kolasinski (1996)—the computer game *Ascent*—involved represented movements primarily in the forward direction only. This suggests that the kinematics of the task performed in the VE may have an important effect on the occurrence of ataxia. For example, VR applications involving limited represented movement—such as teleoperation or simple games—may pose limited risks of ataxia, whereas applications involving a high degree of represented movement—such as highly dynamic games—may pose greater risks of ataxia. Clearly, this unresolved issue is a critical one that must be investigated further.

Research on simulator sickness in VEs should also look at one area that has been neglected in the military simulator environment. Although studies indicate that sickness can occur, little—if any—research has investigated whether such sickness has an impact on training effectiveness. Given the great emphasis often afforded to the use of VR technology for training and education, investigation of the effects of sickness on training effectiveness is an important research issue whose time has come.

APPLICABLE COMPLEMENTARY EFFORTS

As is clear from the above discussion and the references therein, a plethora of complementary efforts—both past and present research—exist in the area of simulator sickness. Most of these efforts are directed toward military simulators. Leaders in such research include the Systems Effectiveness Division of Essex Corporation and the Spatial Orientation Systems Department at the Naval Aeromedical Research Laboratory (http://www.accel.namrl.navy.mil).

However, as noted, research specific to VEs also must be conducted to address the phenomenon specifically as it occurs in VR systems. VR research is being conducted in many laboratories around the globe, several of which are also interested in the investigation of simulator sickness. Such laboratories include the Human Interface Technology Laboratory at the University of Washington (http://www.hitl.washington.edu) and the Ashton Graybiel Spatial Orientation Laboratory at Brandeis University (http://www.bio.brandeis.edu/pages/faculty/dizio.html). There are also many laboratories in the United Kingdom conducting VR research. The major VR researchers there have established a group known as the UK Virtual Reality Special Interest Group (http://www.crg.cs.nott.ac.uk/ukvrsig/), made up of representatives from both industry and academia, which aims to provide a communications network for all VR researchers and users in the United Kingdom. Some of the member laboratories, such as the Virtual Environment Laboratory at the University of Edinburgh (http://hagg.psy.ed.ac.uk/), also are interested in investigation of the effects of VR exposure.

A final major contributor to the investigation of simulator sickness in VEs is the Simulator Systems Research Unit (SSRU) of the U.S. Army Research Institute (http://www.ari.fed.us/ssru.htm). SSRU is investigating the use of VEs for the training of dismounted infantry (Lampton et al., 1994a) for the ultimate goal of integrating the dismounted soldier into large-scale networked simulations. As part of its research effort, SSRU also is dedicated to investigation of the occurrence of sickness in VEs (Lampton et al., 1994b).

REFERENCES

Baltzley, D.R., R.S. Kennedy, K.S. Berbaum, M.G. Lilienthal, and D.W. Gower. 1989. "The Time Course of Postflight Simulator Sickness Symptoms," *Aviation, Space, and Environmental Medicine* 60(11):1043-1048.

Casali, J.G. 1986. *Vehicular Simulation-induced Sickness, Volume 1: An Overview.* IEOR Technical Report No. 8501, NTSC TR 86-010. Naval Training Systems Center, Orlando, Fla.

Crampton, G. (ed.). 1990. *Motion and Space Sickness.* CRC Press, Boca Raton, Fla.

Crowley, J.S. 1987. "Simulator Sickness: A Problem for Army Aviation," *Aviation, Space, and Environmental Medicine* 58(4):355-357.

Fowlkes, J.E., R.S. Kennedy, L.J. Hettinger, and D.L. Harm. 1993. "Changes in the Dark Focus of Accommodation Associated with Simulator Sickness," *Aviation, Space, and Environmental Medicine* 64(7):612-618.

Kellogg, R.S., and K.K. Gillingham. 1986. "United States Air Force Experience with Simulator Sickness, Research and Training," in *Proceedings of the 30th Annual Meeting of the Human Factors Society* 1:427-429.

Kennedy, R.S., and L.H. Frank. 1983. "A Review of Motion Sickness with Special Reference to Simulator Sickness," paper presented at the National Research Council Committee on Human Factors Workshop on Simulator Sickness, September 26-28, Naval Post-Graduate School, Monterey, Calif.

Kennedy, R.S., J.E. Fowlkes, and M.G. Lilienthal. 1993. "Postural and Performance Changes Following Exposures to Flight Simulators," *Aviation, Space, and Environmental Medicine* 64(10):912-920.

Kennedy, R.S., M.B. Jones, K.M. Stanney, A.D. Ritter, and J.M. Drexler. 1996. "Human Factors Safety Testing for Virtual Environment Mission-operation Training," Contract No. NAS9-19482.

Kennedy, R.S., D.S. Lanham, J.M. Drexler, and M.G. Lilienthal. 1995. "A Method for Certification That Aftereffects of Virtual Reality Exposures Have Dissipated: Preliminary Findings," pp. 263-270 in A.C. Bittner and P.C. Champney (eds.), *Advances in Industrial Ergonomics Safety VII*. Taylor and Francis, London.

Kennedy, R.S., M.G. Lilienthal, K.S. Berbaum, D.R. Baltzley, and M.E. McCauley. 1989. "Simulator Sickness in U.S. Navy Flight Simulators," *Aviation, Space, and Environmental Medicine* 60(1):10-16.

Knerr, B.W., D.R. Lampton, J.P. Bliss, J.M. Moshell, and B.S. Blau. 1993. "Human Performance in Virtual Environments: Initial Experiments," *Proceedings of the 29th International Applied Military Psychology Symposium*. Wolfson College, Cambridge, U.K.

Kolasinski, E.M. 1995. "Simulator Sickness in Virtual Environments," ARI Technical Report 1027. U.S. Army Research Institute for the Behavioral and Social Sciences, Alexandria, Va.; available on-line at www.ari.fed.us/ssru.htm. Also available by anonymous ftp at ftp.hitl.washington.edu/pub/scivw/publications/SimSick.rtf.

Kolasinski, E.M. 1996. "Prediction of Simulator Sickness in a Virtual Environment," *Dissertation Abstracts International*, 57-03. University Microfilms No. 96-21485. Available on-line at http://www.hitl.washington.edu/projects/vestibular/kolasinski/.

Lampton, D.R., B.W. Knerr, S.L. Goldberg, J.P. Bliss, J.M. Moshell, and B.S. Blau. 1994a. "The Virtual Environment Performance Assessment Battery (VEPAB): Development and Evaluation," *Presence* 3(2):145-157.

Lampton, D.R., E.M. Kolasinski, B.W. Knerr, J.P. Bliss, J.H. Bailey, and B.G. Witmer. 1994b. "Side Effects and Aftereffects of Immersion in Virtual Environments," *Proceedings of the 38st Annual Meeting of the Human Factors and Ergonomics Society*, Vol. 2, pp. 1154-1157.

Mon-Williams, M.A., J.P. Wann, S.K. Rushton, and R. Ackerley. 1993. "Real Problems with Virtual Worlds," *Ophthalmic and Physiological Optics* 13:435-436.

Reason, J.T., and J.J. Brand. 1975. *Motion Sickness*. Academic Press, London.

Regan, E.C. 1993. "Side-effects of Immersion Virtual Reality," paper presented at the *International Applied Military Psychology Symposium*, July 26-29.

Regan, E.C., and K.R. Price. 1994. "The Frequency of Occurrence and Severity of Side-effects of Immersion Virtual Reality," *Aviation, Space, and Environmental Medicine* 65(6):527-530.

Rolland, J.P., F.A. Biocca, T. Barlow, and A. Kancherla. 1995. "Quantification of Adaptation to Virtual-eye Location in See-thru Head-mounted Displays," *Proceedings of the Virtual Reality Annual International Symposium 1995*. IEEE Computer Society Press, Los Alamitos, Calif., pp. 56-66.

JOHN N. LATTA
Flights of Fantasy: An Oxymoron—Defense and Entertainment

The lure is engrossing—incredible defense technology being converted to the best entertainment this side of watching war on CNN. Visions of long lines of want-to-be war fighters can be seen making entertainment operators salivate at the thought of bulging bank accounts based on skyrocketing cash flow per square foot. Fantasy or a potential winner? Just a dream. Entertainment is a business, and war fighting is about execution in combat. There is no congruence in commercial business models and military mission statements. Out-of-home entertainment is a social experience, while winning on the battlefield is about doctrine, planning, leadership, and team effectiveness. Defense is also about leveraging technology to superior advantage in war. Yet in entertainment, technology is a lever to increase play rates and draw in the context of social environment. The often-heard chorus is that defense technology has applications in many sectors and the entertainment industry may be one. Yet, for example, in three-dimensional technology the conversion has largely taken place and the fuel of innovation is not Department of Defense (DOD) reuse but entrepreneurs seeking to get rich as they spend venture capitalists' money in new start-ups. DOD can help the entertainment industry by having more movie theaters on military bases.

WILLIAM K. McQUAY

Advances in software and computer technology are making possible complex simulations based on affordable and reusable modeling components. Businesses will soon be able to realize increases in productivity through the widespread employment of simulations as aids for decision making and training. As a result, the commercial marketplace will increase for generic simulation techniques, simulation infrastructure, and off-the-shelf components for applications in financial industries, manufacturing, industrial process control, biotechnology, health care, communication and information systems, and entertainment.

DOD TECHNOLOGY FOR INDUSTRY

The entertainment industry has brought simulation technology and synthetic environments into the media mainstream. However, development of the software to enable such simulations is a manpower-intensive endeavor and thus is costly. Industry has the opportunity to exploit current U.S. Department of Defense (DOD) research and simulation technologies to bring products to market faster and at lower cost. Industry can leverage DOD joint standards and modeling and simulation (M&S) initiatives such as the DOD High-level Architecture, distributed interactive simulation (DIS), joint simulation system (JSIMS), the joint warfare simulation (JWARS), and the joint modeling and simulation system (JMASS). The joint M&S standards provide execution frameworks and emphasize models based on interoperability, reuse, portability, distributed operation, scalability, broad applicability, technological evolvability, and maximum feasible use of commercial off-the-shelf software. A potential high-payoff defense simulation technology is desktop M&S—simulation brought to the personal computer on the desktop of the engineer, analyst, and decision maker. Desktop M&S technology could be the basis for future video games, Internet games, or location-based attractions.

As entertainment simulations become increasing complex, the industry will face some of the same challenges faced by DOD in military simulations. As a result, DOD and industry could benefit from technology sharing in such areas as:

- Extensible architectural frameworks for tools and models that support a "plug-and-play" concept;
- The ability to geographically distribute simulations across a heterogeneous computer network;

- Simulation development tools to support creation of model components that comply with architectural standards;
- Multiple language support: a user can specify the target source language (C, C++, Objective C, Ada, Java, etc.) to ease the transition to Internet-based entertainment; and
- Object-based technologies to allow component reuse in different products and on different platforms.

COMMERCIAL TECHNOLOGY FOR DOD

It is current DOD policy to use commercial off-the-shelf software whenever it meets DOD requirements. The DOD joint standards are designed as open systems architectures that support commercial off-the-shelf software and tools. The commercial sector has been very successful in developing two- and three-dimensional visualization software and in creating virtual reality applications. Such tools are more affordably and efficiently created by industry and can be maintained at low cost by a broad customer base.

Under a collaborative M&S marketplace concept, industry could build commercial and entertainment simulations based on DOD frameworks and reusable components and supplement them with advanced visualization technology and animation. DOD could employ these commercial products as needed to meet individual organizational requirements. Broad DOD and military service requirements could be satisfied by core joint M&S and supplemented by multiple commercial tools and capabilities from the collaborative M&S marketplace. DOD has insufficient resources to purchase DOD-wide licenses for the multiplicity of unique and individual products required for all DOD and service organizations. Instead, the collaborative M&S marketplace becomes a new outlet for commercial application developers where the DOD field organizations buy the exact product they need. Companies will have a new arena for sales of commercial products (tools and eventually even model parts) compatible with DOD joint standards. The best of DOD and commercial technology would be available to both sectors.

JACQUELYN FORD MORIE
The Military and Entertainment: Historical Approaches and Common Ground

The military and the entertainment industries have come to their respective uses of technology from very different directions and motivations. The military has typically started with an existing need: for training people how to fly an airplane, for example, or for better communications. The military has then been extremely successful in creating the technology that will meet those needs—thus producing the better-trained, or better-informed, individual. The creation of a technology is driven by need. The entertainment industry, on the other hand, has typically started with existing technology but has been very good at creating a need within the audience that will bring the people in—to the arcade, the theme park, or other venue. The need, be it for an experience that continues the story of a popular film or a way to move people around a park, comes after the technology that supports it.

It is immediately apparent that there is a great deal of common ground in these two approaches. The military and entertainment industries have been complementary for longer than one might realize. There is a sign on an airplane simulator invented by Edwin Link in 1930 at the U.S. Air Force Armament Museum in Pensacola, Florida, that states that it was originally designed as an entertainment device. This "Blue Box" was sold to amusement parks until 1934, when Link, a pilot himself, met with the Army Air Corp to sell the Corp on the concept of pilot training with his device. The rest is history. The key here is that people enjoy interesting and satisfying experiences, whether for job enhancement or personal enrichment. For the military, the experiences provided by the technology were directly applicable to better performance in the mission of the job. They worked because they were interesting and pleasurable, as well as realistic, ways of learning the task at hand. For entertainment audiences, the motivation is more self-centered and aimed at enhancing one's personal time. These "civilian" experiences are motivated by several desires: thrill seeking, escape from one's everyday world, social interaction, or self-betterment (physical or mental).

What the military did in accepting Edwin Link's idea to use his entertainment device as a trainer has been echoed in more recent times by the appropriation of military technology by the entertainment industry. The common ground is an invention or an idea that lends itself to multiple uses. Where do these ideas comes from? Many of them come from a fertile environment for thinking and creating. For years the military has utilized just this kind of environment within the academic walls of university research labs to help develop some of its more cutting-edge ideas. By investing in these groups, the military has allowed ideas to ferment in diverse locations with heterogeneous teams of people. Over the decades, it has received a very nice return on its investments. Until recently, however, very few entertainment companies had taken advantage of the potential of these same research settings.

In 1991 I proposed to my research laboratory, the Institute for Simulation and Training (IST) at the University of Central Florida, a new initiative designed to bring together entertainment companies with what I saw as the related research we were doing for the government in virtual reality technology. Working with a theme park design professional, Chris Stapleton, as my partner to determine areas of common interest, we set about to bring the entertainment industry to a working familiarity with the latest in digital research, in a project we dubbed "Operation Entertainment." Dozens of entertainment professionals came to IST over the next three years; we brought them in for endless demonstrations of what we were doing and intense discussions of how the work could apply to their profession. While we were never able to convince them to invest money in our laboratory, there were many seeds planted and several successes. One was when we advised Doug Trumbull on computer technology and connected him with an Orlando business from which he purchased the equipment to start up his company to produce the Luxor project. The second was in the creation of "Toy Scouts," which are discussed in the following paragraph. Many of the ideas developed through the history of Operation Entertainment have pointed the way to where the entertainment industry might go if it were to invest in the research labs that are already out there. Japanese companies have been doing so with the largest labs and the entertainment giants are starting to follow suit. There are many more labs out there as well that could prove extremely useful as the technology develops, and since many of them are already involved in military research as well, there is a great potential to maximize this research so that it benefits both groups. This is truly the best and most promising common ground. But exactly how can this type of collaboration be accomplished? Entertainment companies certainly don't have the dollars to invest in research the way the government does. This is, for the most part, true, but there are new ways we can think about collaboration and mutual discoveries.

One example is embodied in my work with a group at the Institute for Simulation and Training's Visual Research Laboratory that we called the "Toy Scouts." This was a group of undergraduate computer science and art students who met on Friday nights to see what they could do with the treasure chest of military "toys" that existed in our research laboratory. Guided by volunteer researchers in the lab, and with the outside advice of some local entertainment experts who would periodically visit, the students developed truly innovative full-body immersive games using virtual reality technology. One of the games was called "Nose Ball." In Nose Ball you used your nose as the paddle that controlled the ball in a three-dimensional breakout game. Because it was in the center of your stereoscopic vision, it was a perfect aiming device. Nose Ball was also a full-body physical workout. In the four years of the Scout activity, approximately a dozen new full-body immersive games were developed, with many clever and innovative ways to interface with the technology. These students, with their raw energy and fresh approaches, came up with ideas that might not have occurred to the more seasoned professional. The students benefited educationally from the expertise of the researchers they worked next to, and the researchers were often able to look at things with fresh eyes because of their close proximity to the Scouts. The entertainment industry was able to get new ideas from this work, and it became a wonderfully synergistic approach and experience to all involved.

The military has long partnered with the academic research community as an integral part of the discovery and implementation process for bringing new technology and techniques to a state of usefulness. The above example of the Toy Scouts is only one suggestion of how the military and entertainment industries can find common ground in academic research laboratories. The entertainment industry could sponsor such groups around the country at military research laboratories, and both groups could reap the rewards. No doubt there are many more ways that can be imagined; if only a fraction of them are implemented, the benefits might amaze us all.

JACQUELYN FORD MORIE
The Evolution of Entertainment: Who's in Charge?

In the entertainment realm the audience is starting to become more and more sophisticated. Reversing a decades-old decline that has continually devolved an audience into ever-more-passive beings, today's audiences are eager and hungry for more direct participation. Fueled partly by home video games, and partly by the Internet, participants want more and more control over the experiences they are being offered. Video games appeal because the player is in control; one achieves a sense of satisfaction by reaching ever higher levels at one's own pace. The Internet is engaging in large part because it empowers the user to be a producer as well as a consumer. The entertainment industry, by contrast, driven as it is by economics of throughput and ticket prices, wants neither producers or controllers as its perfect audience. A passive audience allows for the most control over the numbers and timing of the attractions. However, the result of this is boredom: while the attractions grow ever-more grandiose and able to accommodate ever-larger crowds, the audience tires quickly and does not come back for repeated plays. The people do not feel themselves an active part of the experience. The audience has the ultimate control—it speaks with its time and its wallet. The entertainment industry will find it more difficult to continue in the old proven formulas of canned events that an audience is driven, flown, walked, or bumped through.

The next decade will see a trend toward what audiences demand—more control and empowerment. This will happen in several ways. The first is through more individual and unique play experiences, the second through more team play experiences, and the third through more spectator experiences. A few words on each are in order.

INDIVIDUAL PLAY EXPERIENCES

Individual play experiences appeal to our need for a self-directed experience, even if done in a social setting. They need to progress beyond individual home or arcade video games and extend the level of interactivity far beyond simple repetitious button punching.

This area was one I worked in for several years at the Institute for Simulation and Training's Visual Research Laboratory with a group we called the "Toy Scouts." This was a group of undergraduate computer science and art students who met on Friday nights to see what they could do with the treasure chest of military "toys" that existed in our research laboratory. Guided by volunteer researchers at the lab and with the outside advice of some local entertainment experts, these students developed truly innovative full-body immersive games using virtual reality technology. One example was a game called "Nose Ball." In Nose Ball you used your nose as the paddle that controlled the ball in a three-dimensional breakout game. Because it was in the center of your stereoscopic vision, it was a perfect aiming device. Nose Ball was also a full-body physical workout. In the four years of the Scout activity, approximately a dozen new full-body immersive games were developed. It was far cry from the couch potato mentality we might have expected from the video game and TV generation. In fact, this is an innovative way to combine sports and simulated experiences—a wonderful athletic hybrid. Think of going to some future digital gym for a Nose Ball workout!

While immensely popular with the audiences who experienced them, the drawback to games for the entertainment industry is economics. The games were so enjoyable that the typical experience was 10 to 15 minutes long. Add to that the suiting up time and lead-in of how to play, and there just couldn't be enough return on an investment to make a profit. For this to evolve, the technology needs to be cheaper and easier to use, but it also requires a new way of thinking about technology as something active, vibrant, and participatory, with innovative interfaces that extend interactivity far beyond simple button pushing.

TEAM PLAY EXPERIENCES

A second big challenge for entertainment companies today is how to make computer interactivity play to a group larger than just a few people at a time. The military solved this problem years ago with SIMNET. As the grandfather of this area, SIMNET provided not so much prescribed scenarios but a common ground for participants to work together toward a goal. We have seen only a handful of successes in the entertainment community so far, and these involve fairly small-sized audiences—typically 12 to perhaps 100 people.

There is definite need to continue to develop experiences in this realm. These types of activities fulfill our need as social beings to work together and communicate with one another in a group situation. This is one of the reasons why Internet chat groups are so popular. The best and most successful group experience to date, especially in terms of the larger audience, is Loren Carpenter's 1991 interactive piece shown at SIGGRAPH in Las Vegas (and again at SIGGRAPH 1994 in Orlando).

Loren's "game" not only allowed for 3,000 to 5,000 simultaneous players to control a "pong" game or a flight simulator, but it did so while building a level of group excitement and involvement that has rarely been seen in our current digital entertainments. A surprising outcome of this game was that the audience as a whole did not perform at an average level, as might be expected, but at a much higher collective performance level. What heightened the level of the collective fervor was that the individual audience members could immediately sense their influence on the outcome. More work needs to be done at this level of team play.

SPECTATOR EXPERIENCES

An obvious extension to the realm of team play is that of spectator play. Not everyone involved with digital entertainment will want to be a direct participant. Sometimes people enjoy themselves when they are engaged as a spectator. Being a spectator is not necessarily about being passive; it is about being a participant with anonymity within a crowd. This provides some people a less threatening forum in which to express themselves. Look at football or other team sports as the best example: only a small percentage of the participants actually play. The bulk of the industry (as well as the money to pay the players) is built around the fans. There is a potentially huge market to be developed for providing a substantial and rewarding spectator experience in the digital entertainment realm. So far no one is exploring this avenue.

These types of experiences require a new collaboration of entertainment with its audience. The military, in this respect, has been most responsive to its audience—not only the individual player but also the group dynamics that it served to train or connect. The thing to remember is that technology itself will not sell anything beyond a momentary novelty. It is the larger experience that will spell success or failure, and it is in giving the audience what it desires that the most successes will be found. It is up to us to find the ways to do this.

JEFFREY POTTER

Our segment of the computer graphics market extends from the plug-in card for the home personal computer all the way up to the high-powered workstation graphics accelerator for engineering industrial use. We expect to see the natural increase in renderer horsepower and on-line storage capabilities that the computer industry has become accustomed to. Every 12 to 18 months, the processing sees about a twofold increase in performance, with storage capacities moving at nearly the same pace. Simultaneously, we expect to see features once reserved only for the expensive workstation market to gradually filter down and become available to the home computer user. These features include high-quality antialiasing, acceleration of both geometry and display processing, and advanced texturing capabilities. Simultaneously, we expect to see new exotic ways in which three-dimensional (3D) computer graphics can be applied to the common tasks done in a 2D world today. Remember, not too long ago we were using 24-line, 80-character, alphanumeric-only displays to do our word processing and spreadsheets. With the advent of inexpensive 3D graphics, ordinary 2D graphics might seem quaint and backwards in just a few more years.

Like any product that undergoes evolutionary change, computer graphics products will react to developers' needs. Operations that become the most commonly used routines performed by the host central processing unit (CPU) in software will eventually migrate to hardware. The host CPU is then able to control rendering at a higher level, and developers can start thinking up the next big processor-intensive algorithm. We do not see a fixed set of features being used to separate the personal computer (PC) market from the workstation market. The line between personal computer graphics and workstation graphics will be more rooted in price points, not capabilities. That is to say, what we consider to be workstation-quality graphics today will be on every PC owner's desktop in a couple of years. Of course, what will be on the workstation at that time will be limited only by our imagination today.

The enabling technological advances are primarily what has driven the computing industry so far:

- Semiconductor process and geometry—the push to fit ever more gates onto reasonably priced pieces of silicon while keeping thermal and mechanical problems under control. This matters to both the "number crunching" hardware and the random access memory.
- Memory bandwidth—developing newer higher-bandwidth memory architectures that adapt readily to the 3D graphics paradigm.

- Interface standards—such as the advanced graphics port, allowing the processors and custom-rendering hardware the capability to take advantage of new higher-bandwidth memory.
- New algorithm development—especially in areas such as image compression to further enhance the apparent processing speed of a system.

The research challenges are to invent the next "big thing" in computer graphics. Our Compu-Scene IV product practically stole the market in high-end military flight simulation and training in 1984 when we introduced photographic-quality texturing to real-time graphics. Research and development must strike a happy medium between finding the next gee-whiz feature that engineering can dream up and the marketable improvements that translate into increased sales.

In our experience, one market drives the other, and occasionally developments and feature sets come full circle. U.S. Department of Defense (DOD) applications concentrate on real-world accuracy and training effectiveness. Entertainment applications want the "look and feel" of the high-powered military simulations but at consumer price points. So the products for the entertainment market are designed with carefully chosen compromises based on engineering/marketing research and user feedback. These commercial products then sometimes catch the interest of military customers, who realize that some lower-fidelity systems (such as part-task trainers) can deliver effective training with these compromises.

The drive to create interactive entertainment over the Internet is a prime example of complementary efforts. The lessons learned by the defense industry suppliers involved in the Distributed Interactive Simulation standard can be put to good use by the entertainment community.

We have had a close working relationship with Sega Enterprises, Ltd., developing the graphics hardware systems for the Model 2 and Model 3 arcade systems. This drove us to miniaturize our image generator architecture and to develop new algorithms for such features as antialiasing. We have used this cross-pollination of ideas to enhance our product line, most notably the R3D/100 chop set and R3D/PRO-1000 system. The R3D/PRO-1000 system is then able to serve markets that previously required expensive workstation-based systems at lower cost.

DAVID R. PRATT
Military Entertainment?

The Joint Simulation System (JSIMS) is the flagship program of the next generation of constructive models. JSIMS is a single, seamlessly integrated simulation environment that includes a core infrastructure and mission space objects, both maintained in a common repository. These can be composed to create a simulation capability to support joint or service training, rehearsal, or education objectives. JSIMS must facilitate Joint Service training, significantly reduce exercise support resources, and allow user interactions via real-world command, control, communication, computing, and intelligence (C^4I) systems. The final system will support the ability to resolve down to the platform level the development of doctrine and tactics, mission rehearsal, linkages with other models (e.g., analytical, live, virtual), and a wide range of military operations.

As outlined above, the modeling and simulation (M&S) goals of JSIMS are undoubtedly bold and ambitious. Early on, service- and agency-specific programs were identified to be part of the overall JSIMS program. Based on the three pillars of the Defense Modeling and Simulation Office's common technical framework (conceptual model of the mission space (CMMS), High-level Architecture (HLA), and data standards) along with technology infusion provided by Defense Advanced Research Projects Agency programs (such as the Synthetic Theater of War and Advanced Simulation Technology Thrust), JSIMS represents the first true U.S. Department of Defense (DOD) community-wide M&S developmental effort. The question is whether JSIMS can possibly leverage off of M&S efforts from outside DOD, in particular those from the entertainment industry. Foremost, the goals for a successful military simulation and an entertainment simulation are markedly different. In entertainment the driving factors are excitement and fun. Users must want to spend their money to use it again and again (either at home or at an entertainment center) and hopefully be willing to tell others about it. Unrealistically dangerous situations, exaggerated hazardous environments, and multiple lives and heroics are acceptable, even desirable, to increase the thrill factor. On the other hand, defense simulations overwhelmingly stress realistic environments and engagement situations. The interactions are quite serious in nature, can crucially depend on terrain features or other environmental phenomena, and generally rely on the ability to coordinate jointly with other players. The value of these defense simulations is measured in terms of training and insights revealed. A successful military simulator could be deemed boring and therefore useless in terms of entertainment. Similarly, a successful entertainment simulator could be deemed unrealistic and therefore useless in terms of military training. However, I believe there exists a potential for DOD and the entertainment industry to leverage off each other's M&S efforts provided there is an understanding of how the two fundamentally differ and what each strives to do best.

From an operational point of view, there are three hard technological challenges facing JSIMS: synthetic environment (SE), computer-generated forces (CGFs), and resource reduction. To gain a level of confidence in the outcome of defense models, the models must realistically and consistently represent all of the battlespace in the SE. Tactically significant interactions with the SE, such as rain affecting mobility and line of sight, cross-environment interactions so that objects from the air domain can engage objects from the land domain seamlessly, must be simulated realistically across multiple types of platforms with different underlying terrain representations. CGF behaviors of entities in the simulation need to be flexible and rapidly configurable by end users, and the generated behaviors must continue to evolve through the experience gained as part of the exercises much like humans do in battle. Resources (in terms of time, equipment, and personnel) that currently drive training schedules must be reduced from their current levels. It simply takes too much to set up a simulation exercise. The goal is 96 hours versus the current six months.

The large-scale joint service nature and complexity of JSIMS generally preclude it from taking advantage of using much of the SE framework developed by the entertainment industry so far. However, efforts in the development of user interfaces, use of avatars, and artificial intelligence are of potential interest. User interface development is largely driven by the entertainment industry already as it is the primary means by which its customers experience the desired thrills. The defense training community could benefit from immersive user interfaces that permit more realistic interactions with the SE. Also of interest are more natural interfaces to effectively manipulate large numbers of CGFs or some aspect of the SE, as are the use of avatars to convey information. M&S-driven computer technology advancements that result in the availability of cheaper hardware to do complex computations efficiently, increased personnel expertise, and improved user interfaces could contribute to a significant reduction of resources required to conduct a simulation exercise. Artificial intelligence in CGFs used to populate environments of both defense and entertainment simulations can likely be leveraged provided that they can be flexibly programmed to carry out a variety of tasks and can exhibit advanced behaviors such as the capability to learn. This is the current challenge facing the CGF community within DOD today, and I pose it to the entertainment industry as well in hopes that we may be able to work together on this difficult problem. I have not been able to find a technical reason why the defense and the entertainment M&S communities cannot leverage off each other's efforts. A cross-pollination of ideas between the two appears fruitful provided that their differing M&S goals are not adversely compromised. In general, negative military training, which could result from lack of simulation fidelity or ambiguity in a user interface, is considered to be worse than no training at all.

ALEX SEIDEN
Electronic Storytelling and Human Immersion

The past half-decade has seen a renaissance in digital effects in motion pictures. Correspondingly, the use of certain "traditional" effects technologies, such as compositing with optical printers, has diminished greatly. Writers and directors have been given a new and powerful set of tools to realize their visions. New techniques have made the impossible possible and the prohibitively expensive more affordable. Additionally, a tremendous amount of effects work is in the "invisible" category: wire and rig removals, sky and background enhancements, and so on.

Box office success fuels much in the world of filmmaking. (I am not so cynical as to say it is the only force in operation.) The tremendous returns on *Terminator 2: Judgment Day* and *Jurassic Park* exploded studio interest in visual effects and the facilities that create them. Many studios have made substantial investments in their own effects units. Currently, films such as *Twister* and *Independence Day* reinforce this trend. The demand for visual effects has never been as high as it is today, and it will continue to grow for the next few years. After leveling off, there will doubtless be a shakeout in the number of facilities providing these services; in some ways this may have already begun.

Visual effects and simulation computer graphics in the visual effects world have very little to do with "simulation," as the term is commonly used in the computing community. The broadest definition of a simulation is any synthetic or counterfeit creation. However, most in this audience would consider a simulation as being a mathematical or algorithmic model, combined with a set of initial conditions, that allows prediction and visualization as time unfolds. This generalized model allows easy manipulation of the initial conditions and parameters, such that many possibilities can be explored.

Computer graphics in visual effects has more to do with what is euphemistically called "hybrid" or "empirical" techniques, and more candidly called "grotesque hacks." The dinosaurs of *Jurassic Park* were not "simulated" any more than pre-World War II Los Angeles was "simulated" for *Chinatown*. Certainly, any shot in the latter film has a rich sense of place and time: the office of private investigator Jake Gittes is powerfully evoked, and the audience imagines the rest of that world. But turn the camera 30 degrees and you'll see a bunch of C-stands and some grips. Changing views requires substantial time—though the illusion of a complete and continuous world still exists. Similarly, changing the action of a synthetic creature, such as a dinosaur, requires extensive rework. No single generalized conceptual model exists for those dinosaurs and their important visual properties, such as their gait cycles, the movement of flesh over bones, or the texture of their saurian hides. Often, a precise simulation would not only be more complicated but would also be aesthetically undesirable; for example, the scale of dinosaurs in *Jurassic Park* changes dramatically from shot to shot and sequence to sequence.

TECHNOLOGICAL ADVANCES AND RESEARCH CHALLENGES

Turnkey animation systems have advanced greatly in the past several years. Such advancements include flexible inverse kinematics that make creature animation practical, particle system front ends that allow complicated dynamic effects, and the big strides in software on Macintosh and personal computer-compatible systems. Nevertheless, software is the single largest area where attention should be focused. Animators and technical directors face daunting challenges as shots become more and more complicated. Techniques must be developed that allow more facile management of this increasing complexity. Visual programming, to name just one possible solution, has been proposed as a method of allowing people to work at higher levels of abstraction without sacrificing precise control; other ideas will hopefully emerge as well. Ask the production manager of any effects studio and he or she will tell you the biggest problem is the shortage of skilled animators and technical directors. As such, training and education will be a huge issue for the next several years.

Current renderers lack good simulation of area light sources and diffuse-diffuse interactions. This is critical for matching live-action photography, especially for daylight exteriors. Current solutions rely on difficult, poor-looking substitutes. I expect to see radiosity-based techniques become more common. Motion blur and programmable shading are necessary. Some researchers and industry luminaries have talked with great excitement about the potential for photo-realistic artificial actors. This will certainly see some application, but the interest in a revived Marilyn Monroe or Elvis is perhaps overrated. Hardware will continue its inevitable increase in performance and will be matched step for step by growing computing demands. In the long run, frame-rendering times never fall: my average frame time in 1985 was 20 minutes per TV-res frame; now it's around 30 minutes. The vast amounts of data contained in film-res images place great strain on networks; increasing speed and reliability will help speed production. The continuing development of lossless compression techniques also helps. Hopefully, display technologies will develop that can achieve the quality of film, especially more dynamic range and wider color gamuts. Affordable high-resolution real-time playback devices, such as high-definition television, may be part of the solution.

LINEAR AND NONLINEAR STORYTELLING

It is important not to lose sight of the goal of movies or any entertainment: to expand and enrich our lives, to extend our experiences beyond those we have lived through ourselves, to illuminate and explain the experiences that we have lived, and to do all of this in a way that is engaging and compelling. Any art, particularly film, succeeds when the audience forgets itself and is transported into another world. Visual effects must always serve the story. I say this even though I am fully aware that recent effects-laden box office extravaganzas have been disturbingly lacking in this area. Nevertheless, I don't think many people outside the industry would pay $7.50 to sit through the effects reels (an effects reel is an edited piece of film showing only the shots that have visual effects and omitting the rest of the film) of any of these movies, stunning technical and creative achievements though they might be. In any event, many effects-oriented movies fail both commercially and aesthetically.

The Web, CD-ROM, virtual reality, and other multimedia techniques have been touted as ways to transcend the limits of linear storytelling, to give the audience a richer participation in an imaginary world. I will risk alienating many at this workshop by staking out what I expect to be some unpopular ground: I've never seen a CD-ROM that moved me the way a powerful film has. I've never visited a Web page with great emotional impact. I contend that linear narrative is the fundamental art form of humankind: the novel, the play, the film, even the orally told joke or story—these are the forms that define our cultural experience. Although "interactive" and "nonlinear" forms exist, they have never been paramount in our consciousness. Remember when, around the campfire during summer camp, one person would begin a story, tell it for a few minutes, and then another person would take over, developing the story in their own direction? Similar forms exist in improvisational theater games. And although these techniques can be amusing in small doses, I would argue that they are the exception that proves the rule: nonlinear storytelling forms can exist, but they will never supplant the complete control of the linear storyteller. Now, any new medium—film, radio, and television are good examples—has its curmudgeonly detractors. The infant film industry received similar criticism from those who argued that theater would always be the dominant medium. The critics lacked the vision of what film could and would become, as the language of film developed. Perhaps I, too, lack such vision and in 10 or 20 years will be eating my words. Will the world of interactive nonlinear entertainment grow? Yes, certainly it will. Will exciting and emotionally powerful images be created in these formats? Yes, there are too many talented people and too much money being sent in that direction for nothing to happen. And certainly the Web will become the preeminent forum for the exchange of commercial and scientific information; its significance will exceed that of the cellular phone, the automated teller machine, the fax machine, and the Home Shopping Network combined. This is not a trivial development. Whether storytelling itself will be fundamentally changed depends on a paradigm shift that I would contend is much larger than for other emerging media. To fully evaluate the likelihood and meaning of such as shift requires a careful distinction between what we think of now as a "story" and what we consider a "game" or "environment." A full appraisal of the differences between the cognitive processes involved is beyond the scope of this paper and is an excellent subject for further research.

STEVEN SEIDENSTICKER
Distributed Simulation: A View from the Future

The battle date is August 17, 1943. I am the ball turret gunner of Luscious Lady, a brand new B-17F of the 427th squadron, 303rd Bombardment Group, of the Eighth Air Force. Our takeoff from Molesworth was without incident, but as soon as we were off the ground the pilot asked me to check the wheels. He had an indication that the left main gear had not retracted fully. I hopped into the ball and spun it until I had a good view of the wheel. It looked OK. We chalked it up to a bad indicator in the cockpit. Although the ball with its twin 50s is primarily intended to protect a B-17 from enemy fighters approaching from below, the view from beneath the aircraft comes in handy for other chores. We climb out and begin a long lazy circle. I keep tabs on and report other squadron aircraft as they join our formation.

We are on our second mission and our first over Germany. Our first mission was to bomb a Luftwaffe airfield near Paris. The target was partly obscured by weather. Opposition was light. A few Me-109s came up to meet us. They were not particularly aggressive or well coordinated. Nevertheless, we lost one of our squadron. I saw Old Ironsides get most of her rudder shot off. The pilot was obviously losing control and chose to abandon his ship. I saw 10 good chutes. The debriefing team called the mission a "milk run." The missions would become much tougher as we gained more experience. We were happy to get this far.

My pilot and copilot are in Milwaukee. The navigator/bombardier is in Montreal. Other crew members are in Seattle, San Jose, Denver, and Green Bay. We cannot see or touch each other, but we communicate via what appears to be a B-17's standard intercom. In fact, we are part of a wide-area high-speed data network that connects all crew stations of all aircraft, both friendly and hostile. I don't know the total number of nodes on this network, but it must be in the thousands. The number of spectators who can tap into the net is in the millions. In addition to our voices, this network carries all the data that our individual crew station simulators need to show other aircraft the terrain over which we fly, the weather, and other elements of our environment. To participate in these missions each of us simply dials into the network at the time scheduled for the mission, gets the standard crew briefing on our screens, and waits for our turn to take off. The pilots, bombardiers, and navigators get a detailed briefing on the target and expected weather. The rest of the crew gets briefed on expected opposition. The briefings are, of course, the same as (or as close as possible to) the original briefings given to the original crews. Like in the original briefings, we can ask questions and get answers.

APPENDIX D

173

Not all the crew stations on Luscious Lady are manned by humans. The waist gunners and the radio operator are computer-generated entities. They do their jobs reasonably well. They even respond to us when we talk to them over the intercom. However, if the conversation strays from simple orders or reports they quickly become confused and start spouting gibberish. Some of the other friendly aircraft on the mission and some of the opposing Luftwaffe fighters have no human crews at all. But it's getting harder to tell who is human and who is computer-generated, because the programmers keep tweaking their behavior algorithms. But my personal feeling is that they will never get to the point where these simulations are totally indistinguishable from real people. I hope they don't.

Over the Channel the pilot gives us the order to test our guns. This is a ritual that ensures that the guns are working and marks the real beginning of the mission for us gunners. From here we are in harm's way. I cock both guns, point to a clear area, and let loose with a short burst. The tracers arc away gracefully. I have managed not to hit anyone else in the formation. To do so is considered very bad form. It also requires the hapless shooter to buy dinner for the shootee's crew at our next annual convention. Of course, the computers that run this whole operation keep track of everything, so there is no arguing or hiding. The target today is the Me-109 plant in Regensburg. We know that the Luftwaffe was out in force that day. The Eighth Air Force lost 24 B-17s out of a force of 147. Shortly after we cross the French coast the nose gunner shouts "four 109s at 12 o'clock low." The control yoke feels comfortable in my hands as I spin the turret forward. They are coming at our formation four abreast from dead ahead. The winking lights on the leading edge of their wings show that they are firing. I mash the right pedal hard to tell the lead computing gun sight to use maximum range. The left pedal goes to the third notch to input the wing span of an Me-109. I line the sight's pipper on the number two plane and fire short bursts, trying to adjust the range as they close. My shots appear low. Just about everyone in our formation is firing. A puff of smoke bursts from the number three fighter. It continues to smoke as their formation passes right through ours.

This line abreast head-on attack was developed by the Luftwaffe in early 1943. It took a lot of courage and discipline on the part of the German pilots, but it was very effective. The idea was not only to get the best shots possible but also to intimidate the bomber pilots and break up the formation. It was probably the greatest game of chicken ever and it frequently ended in collision. The right waist gunner reports another formation at the four o'clock level. But they are out of our range and overtaking us on a parallel course, no doubt moving up for another head-on pass through the bomber stream. I can see their yellow cowlings and know that they belong to JG 26, the "Abbeville Kids," one of the best Luftwaffe fighter wings.

The attacks continue sporadically until we are about 30 miles from the target. At that point we start seeing the dreaded flak. The small black clouds bloom innocently in the distance, but we know that as the ground gunners adjust the aim of their 88s, the bursts will be right around us. There is little evasive action that a formation of B-17s can take. We are near the IP (initial point) that the pilot must fly over if we are to get our bombs anywhere near the target. At that point, the bombardier takes over and actually flies the plane to the bomb release point, using autopilot controls on the famous Norden bomb sight, probably one of the most famous but overrated technical developments of World War II. The flak rounds get closer.

The concussion from one of them is louder than the fifties going off next to my ears. The pilot reports that number four engine is starting to vibrate and that the manifold pressure is dropping. Bad news. If it fails we will have to drop out of the formation. Like the weak separated from the herd, we will be on our own. We may have to fight packs of fighters as we try for the coast and the protection of friendly Spitfires. Most who have been through this say that it can be the most exciting part of an afternoon of simulation, but the B-17 seldom survives. Those that do get an award at the next convention and, of course, their battle with the fighters is replayed on the large screen.

We finally reach the target, the bombardier hits the pickle switch, and I watch the bombs fall away. I loose sight of them after a few seconds, but shortly thereafter see a string of explosions on the ground. The bombs land in a rail yard just east of the target complex. But that's closer than the original crew came in 1943.

The flight back was challenging. For two hours we endured more flak and almost constant fighter harassment. Our pilot managed to coax enough power out of the number four engine to maintain our position in the formation. The rest of the formation was not so lucky. Stric Nine took an 88-mm round in the right wing root and the whole wing came off. There were no chutes. Wallaroo lost an engine and had to drop back, but we were close to the coast and a flight of P-47s escorted her back. Once we got over the Channel I turned over my role to an automatic ball turret simulation and had a quick dinner in the kitchen with my wife. I doubt that the rest of my crew even noticed I was gone. I rejoined the simulation for the debriefing. The colonel told us that we had done reasonably well for a second mission crew.

My ball turret is a medium-priced model from RealSim Inc., one of the rising companies in this field. It provides a lot of fidelity for the price and has a lot of update options. I'm very happy with it. The ball spins and rotates vertically much the way the original did and takes up less than half of my garage. The visual scenes are presented on panels built right into the ball. Sound and vibration are provided by some large but ordinary speakers. RealSim sells the basic turret dirt cheap but knows how sim-heads get hooked on fidelity, and so they offer a large range of add-ons that can become real expensive. Some of my colleagues have mounted their units on electrically driven motion platforms. I don't know if that is worth the extra cost. Maybe next year. Many other simulated crew stations are built around virtual reality goggles. Those are a lot less expensive but work quite well. One enthusiastic crew has built a whole B-17 fuselage in a warehouse.

As in most simulations, visual scenes provide the dominant cues. The simulation industry long ago reached its holy grail of creating visual images that are indistinguishable from the real thing. The processing power needed to create them is so cheap that the image generators are no longer a cost factor in most simulators. Databases that represent the terrain of any portion of the earth are readily available at any resolution desired. Specialty "period" databases (Dunkirk or Waterloo for instance) for groundpounders are becoming available but are very expensive.

The key factor that made this kind of group simulation possible was the development of the DIS (distributed interactive simulation) standards about 25 years ago. Once these standards were in place, the designers and builders of simulator components didn't have to spend any more time thinking about linking them together than does the designer of a railroad car need to worry about how to couple his car to a train. The DIS standards allowed the simulation industry to concentrate on functionality, performance, and cost reduction.

My wife used to ask me why I spend so much time and money on this. There are a number of reasons. I, like most middle-aged guys, have often fantasized about going into battle to test my wits and skill with a comparably equipped enemy. In this fantasy I support my comrades and in turn depend on their support. I yearn to experience the heat of battle, victory over my adversary, or a narrow escape from the reach of his weapons. However, I have no desire to shed any of my blood.

I also love history, great battles in particular. I know of no greater battle than that between the U.S. Eighth Air Force and the German Luftwaffe in 1943 and 1944. The leaders of the American forces felt that they could win the war with heavy bombing of German military and industrial targets. To be accurate this had to be done in daylight. Escort fighters of the day did not have sufficient range to cover the bombers. The bombers had to depend on their own defensive weapons.

Participation in these re-created battles is available at a number of levels. I started as a spectator. The magic carpet mode of my computer let me observe operations from any point in space. It also let me attach myself to any aircraft in the battle and listen to the radio and intercom traffic for that aircraft. Running commentary is available from experts. Previews and schedules of upcoming battles are carried by the major sports pages. Reports of completed battles also are carried. These tend to dwell on the personalities involved and the shoot-em-up aspects. How close the reenactment came to the original battle seems to be getting lost.

After watching several of the major raids, I was hooked and wanted to play an active role. My first desire was to be a Luftwaffe pilot, but the requirement for fluency in German eliminated that. Rumors are that an English-speaking Luftwaffe wing is forming. My second choice was to sit in the cockpit of a B-17. But, like the original aircrews, I needed training. The training course for all pilot positions is long and demanding. I opted for the less ambitious role of gunner. Fortunately, the simulator technology that I own trains me more efficiently and quickly than did similar training programs in 1943. After a few intense weekends, I passed the qualification tests and was assigned to my present crew. We are not the most proficient crew on today's raid, but neither were the new crews in 1943.

As I become more serious in this avocation, I wonder where it is going. Some social commentators are starting to decry the "glorification of war." Others counter with statements about "harmless outlets of male aggression," despite the fact that at last year's convention the Best B-17 Crew Award went to an all-female crew. Some critics are worried that the super-realistic simulation available today is going to replace drugs as the national addiction. Who knows! The raid on the ball-bearing factories in Schweinfurt is scheduled for next week. It was the bloodiest for the Eighth Air Force. I think my crew and I are good enough and lucky enough to survive. I can hardly wait to find out.

JACK THORPE
Research Needs for Synthetic Environments

PURPOSE

This paper introduces one approach for thinking about the technical challenges of constructing synthetic environments and some of the related research issues. The paper is designed to stimulate discussion, not to be a comprehensive treatise on the topic.

DISCUSSION

Simulation, virtual reality, gaming, and film share the common objective of creating a believable artificial world for participants. In this context, believability is less about the specific content of the environment and more about the perception that there exists a world that participants can port themselves into and be active in—that is, exert behavior of some sort.

In film, this is vicarious. In simulation, virtual reality, and gaming it tends to be active, even allowing participants to choose the form for porting into the environment: either as an occupant of a vehicle moving through the environment, as a puppet (proxy) of him or herself that he or she controls from an observation station, or as a fully immersed human. The iconic representation or avatar can assume whatever form is appropriate for the environment.

When the participant is an audience member in a single venue and is neither required to interact overtly with other audience members in the same venue or other connected venues, the issues of large-scale interactivity and distributed locations are minimal. On the other hand, when tens or hundreds of remotely located participants are ported into the same world and begin to interact freely (and unpredictably), as demonstrated in recent advances in distributed interactive simulation, not only are the environments more interesting but the technical challenges are also more difficult. It is likely that these will also be the next-generation commercial application for this technology, and so addressing technical issues is timely.

To design and build these more complex worlds, the following major tasks have been found to be useful classifications of the work needed to be done and the tools required to perform this work, thus leading to the research and development needed to construct the tools. For each of these tasks a few of the research issues are identified, but this is far from a comprehensive treatment:

- Efficient fabrication of the synthetic environment;
- Design and manufacture of affordable porting devices that allow humans to enter and/or interface with these environments;
- Design and management of a worldwide simulation Internet to connect these porting devices in real time;
- Development of computational proxies (algorithms) that accurately mimic the behavior of humans unable to be present;
- Staffing, organization, and management of realistic, validated sentient opponents (or other agents), networked based, for augmenting the world; and
- Development of innovative applications and methodologies for exploiting this unique capability.

Efficient Fabrication of the Synthetic Environment

Artificial worlds are usually three-dimensional spaces whose features are sensed by the participants in multiple modes, almost always visual but possibly auditory, tactile, whole-body motion, infrared, radar, or via a full range of notional sensor or information cues. For each of these modes of interaction, the attributes can be specified in a prebuilt database ahead of time, or calculated in real time, or both.

The challenge is to construct interesting three-dimensional environments efficiently. Cost rises as a function of the size of the space (in some military simulations it can be thousands of square miles of topography), resolution, detail (precision cues needed for interaction), dynamic features (objects that can interact with participants, like doors that can open or buildings that can be razed), and several other factors. As a general observation, the tools needed to efficiently construct large complex environments are lacking, a particularly serious shortfall when fine-tuning environments for specific goals of a simulation or game. Toolsets are quirky and primitive, require substantial training to master, and often prohibit the environment architect from including all of the attributes desired. This is a serious problem, one that seems to get relatively little attention. It is an area that needs continual research and development focus.

Design and Manufacture of Affordable Porting Devices that Allow Humans to Enter and/or Interface with These Environments

The manner in which the human enters the synthetic environment continues to undergo rapid change. Flight simulators are a good example. Twenty years ago a sophisticated flight simulator cost $20 million to

$40 million. Ten years ago technology allowed costs to drop by a factor of 100. Today there has been another one or two orders of magnitude decrease. Further, each new generation is more capable than its more costly predecessor. This drop in cost, with an increase in the richness of the participant's ability to interact with the environment and other people and agents similarly ported there, is especially important as large-scale simulations are constructed—that is, those that might have 50 or more participants (some military simulations have thousands of participants). The cost per participant (cost per seat) can be a limiting factor no matter how rich the interface.

The research issues include the design methodology that leads to good functional specifications for the simulation or game (the work on selective fidelity by Bob Jacobs at Illusion Inc. is relevant), the design and fabrication approaches for full-enclosure simulators (vehicles) and caves (individuals), the porting facade at the desktop workstation (partly manifested by the graphical user interface), and other means of entering the environment, such as while mobile via a wireless personal digital assistant.

Design and Management of a Worldwide Simulation Internet to Connect These Porting Devices in Real Time

Small-scale as well as large-scale distributed interactive environments have baseline requirements for latency, which is compounded when a requirement to worldwide entry into environments is added. Latency is influenced by the type of interaction a participant is involved with in the specific synthetic environment. The requirement is that the perception of "real timeness" is not violated, that is, that participants do not perceive a rift in the time domain (a stutter, momentary freeze, or unnatural delay in consequence of some action that should be a seamless interaction). Because this is a perceptual issue, it is dependent on the nature of the interaction and the participant's expectations.

This becomes a technology issue as the number of independently behaving participants grows, the number of remote sites increases, and the diversity of the types of interactions coming from these sites and participants grows. It has been demonstrated that unfiltered broadcasting of interaction messages ("I am here doing this to you") quickly saturates the ability of every participant to sort through all the incoming messages, the majority of which are irrelevant to a specific participant. The functionality needed in this type of large interactive network is akin to dynamically reconfigurable multicasting, as yet unavailable as a network service.

It could turn out that as the Internet expands it will provide the dedicated protected speed and addressing for these types of interactions, but this is not the case to date, and dedicated networks have had to be installed to support large exercises. Further, it is conceivable that the appetite of the simulation or game designer for more complex and interactive environments will outpace the near-term flexibility and capacity of network providers. Networks are going to have to be smarter, a continuing research issue.

Development of Computational Proxies (Algorithms) That Accurately Mimic the Behavior of Humans Unable to Be Present

Late 1980s experimentation with distributed interactive simulations resulted in the constant pressure to grow the environments in the numbers of participants, but there were never enough porting devices or people to man them to satisfy this growth. Since these environments began as behaviorally rich human-on-human/force-on-force experiences, players demanded that any additional agents brought on via computer algorithm have all the characteristic behaviors of intelligent beings, that is, that they passed the Turing test and would be indistinguishable from real humans—a tall order.

This resulted in a series of developments of semiautomated and fully automated forces capable of behaving as humans and interacting alongside or against other humans ported into the simulation. These developments have met with mixed success. In some cases computer algorithms have been constructed that are excellent mimics of actual individuals and teams, particularly in vehicles, but in other cases the problem is more difficult, especially in areas of mimicking cognition as in decision making. Nonetheless, the commercial application as well as the defense application of large-scale interactive environments will require large-scale synthetic forces behaving correctly. Given that understanding, predicting, and "generating" human behavior transcends simulation and gaming, this will continue to be a major research area.

Staffing, Organization, and Management of Realistic, Validated Sentient Opponents (or Other Agents), Networked Based, for Augmenting the World

Where environments require teams of people acting in concert to augment the synthetic environment for participants, for example, teams of well-trained and commanded competitors, the opportunity presents itself for the establishment of network-based teams. These could be widely remoted themselves, even though they would be perceived as being ported into the synthetic environment at a single location. The challenge of establishing these teams is less technical and more organizational, typical of military operations, except in the case where these teams are required to faithfully portray forces of different backgrounds, languages, and value systems. Technology can assist with real-time language generation and translation. Behaving as someone from a different culture is more difficult.

Development of Innovative Applications and Methodologies for Exploiting This Unique Capability

The capabilities created through the design and instantiation of a synthetic environment can be unprecedented, making conventional applications and methodologies obsolete. This task recognizes that research is needed on how to characterize these new capabilities and systematically exploit them.